Mankind and Deserts 1

Series Editor
Françoise Gaill

Mankind and Deserts 1

Deserts, Aridity, Exploration and Conquests

Edited by

Fernand Joly
Guilhem Bourrié

iSTE

WILEY

First published 2020 in Great Britain and the United States by ISTE Ltd and John Wiley & Sons, Inc.

ISTE Ltd
27-37 St George's Road
London SW19 4EU
UK

www.iste.co.uk

John Wiley & Sons, Inc.
111 River Street
Hoboken, NJ 07030
USA

www.wiley.com

Library of Congress Control Number: 2020943974

British Library Cataloguing-in-Publication Data
A CIP record for this book is available from the British Library
ISBN 978-1-78630-630-2

Contents

Foreword

Mankind and deserts

Fernand Joly[1] departed from this world before he was able to complete this book, through which he had hoped to share his experiences of and passion for deserts.

"Yet another book on deserts!" some might think; another book to add to the numerous publications dedicated to these alien and fascinating worlds.

This book, however, is different from earlier books, as can be seen from its title "**Mankind and Deserts**". It is based on the singular relationships that are formed between humans and the world of the desert – relationships that are unique because they can be traced back to the very origins of humanity. Indeed, it is from the arid Horn of Africa (East Africa) that large migrations began and it is along the deserts, if not within the deserts themselves, that we find the major cradles of burgeoning historical civilizations. This inhospitable world is also associated with great spiritual leaders such as Moses, Jesus, Mohammed and the Buddha, while serving as the backdrop for adventurers and empire builders from Alexander the Great to Genghis Khan, or from the Incans to the Conquistadors in the Andes and Mexico.

What is this universe that is so barren and yet so mesmerizing?

"**All about a word**" was how Fernand Joly introduced his book: "**What is a desert?**" The ambiguity in this word results from the fact that it has been used in different senses across literature and throughout history. For a geographer-writer such as Fernand Joly, the one fact that stood out was that there was no one desert;

1 Professor at the University Paris VII, Denis Diderot, who spent 15 years at the Moroccan Institute of Science in Rabat.

instead there were multiple *deserts*, diverse and varied, ranging from Death Valley to the Kalahari, from the Namib to the Atacama or the Gobi desert. Each of these is a unique landscape, whose uniqueness was born out of its position with respect to the general atmospheric circulation, its geographic location with respect to the sea and to its relief features. And yet, transcending all differences, there is one constraint that binds them all together: **aridity**, defined as a natural physical state characterized by persistent dryness with the corollary of extremely scarce water resources. Both these concepts, **aridity** and **water**, are at the heart of the following chapters. Aridity (Chapter 3) because it "transcends time and takes over space" and water (Volume 2, Chapters 1–3) because it is the essential resource for all life, especially in the desert. Aridity is distinct from "drought", which is simply a "period with insufficient rainfall". Water is seen through the lens of how it appears on land: "wild water", which flows over slopes in an un-channeled manner (Volume 2, Chapter 1), under the impact of violent but brief downpours, and "concentrated waters", i.e. waters "concentrated" into a channel, fed by distant precipitation upstream of the borders of the desert. As can be seen, there is in fact a true **hydrography of the arid world**. Satellite images, among other sources, offer us clear and accurate reproductions of these systems: *fossil hydrographic networks*, the legacy of ancient humid periods, a map of intermittent water bodies (Volume 2, Chapter 2): *playas* and *sabkhas*, permanent lakes with fluctuating shorelines, such as Lake Chad or Lake Eyre, or large allogenous rivers (Volume 2, Chapter 3) that are born outside the desert but travel through the desert, sustaining life, such as the Colorado, the Niger and the Nile, "the first and most remarkable of rivers in the arid world".

The role played by salts in hot deserts is rarely discussed in a systematic manner. Guilhem Bourrié, geochemist and soil scientist at INRAE, has analyzed the origins and nature of these salts and demonstrated how important these salt deposits in the desert are for humans, whether they live off agriculture, livestock or, indeed, the salt trade (Volume 2, Chapter 4).

Chapter 1 of Volume 3, drafted by Joly, was edited after his demise by Yann Callot, a professor at the University Lyon 2 who is a specialist in ergs and dunes. This chapter examines the importance of **wind** in the desert. Wind, sometimes considered to be more emblematic of a desert than even dryness, counts among the earliest dynamics on Earth, an element that humans have not always been able to control. Indeed, this lack of understanding of wind has sometimes had disastrous consequences for certain projects (see the Green Dam in Algeria).

The final chapter in Volume 3, "**Living in the desert**", was taken up by Marc Côte, Professor Emeritus at the University of Provence, who worked as a professor for 20 years at the University of Constantine. He has drawn on his deep knowledge of the land and the people of the Saharan region to present what he calls "The Desert Civilization".

Finally, it must be noted, with great regret, that, since 2010, “geopolitical turbulences have tended to change the fundamentals, to burn away knowledge and to prevent researchers from keeping in touch with this part of space and humanity.”

Acknowledgments

Most of the illustrations were refined by Éliane Leterrier.

Yvette DEWOLF
Honorary Professor at the University Paris VII, Denis Diderot
Paris
August 2020

1

Introduction: The Concept of a Desert

"You cannot get the desert into a book any more than a fisherman can haul up the sea with his nets."
E. Abbey

1.1. All about a word

Desert is one of those familiar yet ambiguous words whose meaning changes depending on people, time and place. It is one of those words whose various meanings can change the very image one has of reality. The personality of the desert is as difficult to capture in everyday language as in the imagination or in scientific research.

The object and the idea, as well as the words to talk about them, exist in humanity's oldest texts: Babylonian, Egyptian, Hebrew, Chinese etc. In Latinate languages, the word can be traced back to 11th Century Latin. The word first described the result (*desertus*: deserted, abandoned) of an act of separation (*deserere*: to desert, to leave). A little later, the word was used to denote a place (*desertum*: desert), an empty or emptied site, uninhabited or depopulated. The various forms this word has taken over the years reflect this ambiguity.

In Medieval times, hermits would retire to the *desert*. The term, in this context, denoted both the isolation from other people as well as the barrenness of the place, the solitude and mysticism of the situation. In the 17th Century, "deserts" chiefly evoked the idea of chosen spots that were distant and discreet, where one could "flee into a desert from the approach of humans". It was a place that was cut off from the world,

Chapter written by Fernand Joly.

voluntarily so, as was the case at the Port-Royal-des-Champs convent[1], or as a result of circumstances, as in the case of the Camisards, French Protestants who lived in an isolated region of France. This connotation of abandonment or exile persists to this day. For example, we say about someone whose words go unheeded that "he is the voice crying in the wilderness" (Isaiah 40:3) and Charles de Gaulle was abandoned politically between 1946 and 1958 in a period that came to be called his "crossing through the desert."[2] From the 18th Century onwards, however, it is the geographic sense of the word that prevails. A *desert* is a seemingly lifeless region, uninhabited, uncultivated, arid (from the Latin *arere*, to burn, to dry) and sterile due to its dryness. The 19th and 20th Centuries, in turn, saw the rise of new forms of deserts: economic and demographic deserts due to rural flight toward industrial and urban areas. Finally, "desert" is used in a psychological sense to talk about an internal state resulting from a sense of deprivation of the heart or mind.

1.2. Arriving at a definition

1.2.1. *What is a desert?*

Different people will have different answers to this question based on their experience and imagination.

When Fénelon[3] exclaimed, "But look! The most beautiful desert you could ever see. Do you not marvel at these streams that fall from the mountains, these steep rocks?", he was, clearly, singing the praises of one of those wild retreats of which his contemporaries would often dream. Modern-day travel agencies celebrate a more truthful picture of a desert – bare landscapes, a sea of sand and five-star bivouacs. The Paris–Dakar Rally racers, whose tracks despoil the Saharan landscape with impunity, bring to television audiences the idea of a threatening desert: devious, enchanting and sometimes deadly! A place demanding courage and renunciation.

Indeed, the Sahara, so close to Europe and featuring in desert tales so often, is what Europeans use as the model for their concept of a *desert*: a vast mineral landscape, no water, no trees, sand, pebbles, nomads and camels. Majestic dunes, picturesque oases, treacherous mirages, outright fabrications and howlers – all these

1 Editor's note: The Port-Royal-des-Champs Convent, south of Paris, was, in the mid-17th Century, home to a strict and ascetic religious order (Jansenism) whose teaching focused on Original Sin and condemned involvement in any "worldly" activity.
2 Translator's note: In French, desert is also used for "wilderness".
3 Editor's note: Fénelon, a 17th Century archbishop of Cambrai – a city in the north of France – was renowned for his eloquence. He was a champion of women's education, an uncommon stance in those times.

feature in most accounts of deserts, such as Théodore Monod's *Méharées*. However, the North American conception of deserts is very different: there is the desolate Death Valley, the high plains in Arizona and Utah with their monoliths, arches, their dry *rios*, their candelabra cacti, their prairies, the Native Americans and their horses. In Central Asia, the deserts include the steppes, in southern Africa you have the dull sands of the Kalahari dotted with trees, and the vast white lakes, the inconstant giant rivers, the interminable shifting of the red dunes of Australia with eucalyptus, aboriginal settlements, sheep and cattle rearers, remote farms, the rabbit-proof fences, the kangaroos and their dry bush that can turn into green pastures in a rainy year.

The *desert* is all this and more. There are populated deserts, as in Mesopotamia and Egypt, or barren, lifeless deserts such as the Tanezrouft and Lut. There are dry deserts, such as the Libyan desert, and foggy deserts such as the Namib or Atacama deserts. And to these we must also add the frozen deserts found in high mountains and the polar regions. In fact, by this definition, the Antarctic is the largest desert in the world.

A *desert* is, above all, a vast region that is empty, dry and outside time. T. Monod described it as "the kingdom of absence". It is a land of surprises, contrasts and oppositions that can occur far apart or close together, sometimes spread out over years, at others separated by barely a few hours. There is the Asian desert, where you broil in summer and freeze in winter, and there is the African desert, where all the days are torrid and all the nights are cold. There is the arid desert, containing opulent oases, while you have a desert enclosed within vegetation. There are deserts that are endlessly flat, as far as the eye can see, and there are plains with rocky islands or bleak mountains. There is the sudden downpour that breaks the monotony of waterless days, and the dry river that swells into a flood. You have the bare desert that transforms, with the rain, into a carpet of flowers. And there is the seemingly uninhabited desert – where as soon as you stop, someone pops up out of nowhere to look at you and to converse.

A *desert* is also a bare spot that is ideal for contemplation and spirituality, where Man finds himself alone in the face of this immensity, silence and beauty. It is the void, where all fundamental questions arise. It is the Biblical antithesis to the Promised Land. It is the site of fervor, where altars are built and sacrifices made, such as the sacrifice of Abraham; the sacred land where divinity approaches humanity, the land of Moses, Jesus, Mohammed and the Buddha. It is the chosen site for renunciation and retreat, from the medieval monks to the 19th Century Father Charles de Foucauld. It is the site of exaltation, the site of gigantic temples, as can be seen in Babylon, Egypt, Syria, Tibet and in the Andes. It was also the backdrop for adventurers and empire-builders, from Alexander the Great to Genghis Khan, and from the Incans to the conquistadors of the Andes and Mexico. It has served as the arena of war from Libya to Iraq, and the testing ground for the atomic bombs. It is fertile ground for prospectors and operators, the salt-miners of the Sahel and the gold-hunters in the Americas, as well as modern-day oil seekers. Finally, deserts have also served as an observatory for the curious, such as Ibn Battuta and Marco Polo, offered new horizons to explorers, such as Caillé and Barth in the Sahara, Prjevalski and Sven Hedin in Asia, Powell in the Colorado and Eyre in Australia, and been used as a research field for scholars, such as Stein in Chinese Turkestan or Monod in Mauritania.

The desert is also a land of legend, exoticism and dreams, brimming over with fantasies and received ideas. Of course, this aspect of it is slowly diminishing as scientific discoveries and analyses progress and with the increasingly widespread sharing of images from these regions. The Romantics turned the desert into a mysterious land, overwhelmed by light and heat, throwing up strange mirages and sandstorms that could bury entire caravans. It was a hellish land of thirst and wind, of silence and death, that was both fascinating and horrifying. In the colonial period, the desert was always considered a redoubtable and inhuman place, the stuff of myth and legend. It was perilous, meant for military glory (like the French Foreign Legion, whose history waxes eloquent on the legionnaire "feeling the hot sand against his skin") or for punishment (the African battalions). Even today, several fictional deserts survive, often harsher than the real ones. Dating back to a time when those who traversed the deserts were travelers, merchants or armies (rather than explorers and tourists), these accounts are often embellished with personal impressions and adventures; many are full of exaggerations or even implausible details and many were romanticized or often deliberately falsified in order to win glory, for political reasons, or to distract the competition. And so, for the layman, the desert was a secret region, hostile and populated by unknown beings who were strange and redoubtable. Many writers, carried away by their own lyrical writing or innovations, helped spread a deformed and enduring vision of the desert. For example, there was the French writer Eugène Fromentin's *A Summer in the Sahara* (which was actually about the Algerian plains) or Pierre Benoît's *The Atlantide*. These dreamlike deserts, "postiches" as they were dubbed by Monod, are a stark contrast to the more realistic and sobering deserts that feature in Westerns and "atmospheric novels" of the modern age, such as the novels about the American West by T. Hillerman and E. Abbey, or A. Upfield's Australian novels.

1.2.2. *Conceptual deserts: deserts that have been experienced*

Each desert, considered as an object of study, may be dissected, qualified and quantified. However, this does not result in a homogeneous and unique scientific concept of deserts. A *desert* cannot be reduced to a concrete, simple and universal formula. To each their own desert, which seems to them more "desert-like and true" than another's concept. There are, therefore, physical deserts that are objective and measurable by a certain degree of aridity, and then there are psychic deserts, subjective and experienced as a sensation of emptiness and the absence (or, at least, the apparent absence) of human life, and the "true desert" is, in fact, a combination of these two. For Monod, the "true" desert is the "experienced" desert, which one has walked oneself, lived through, feared and admired.

1.3. The world of deserts

In these conditions, it is futile to try and establish a clear definition of a desert that would be applicable in all cases. However, transecting a desert (the Sahara, for instance) from its periphery to its center (or the other way around) could help in revealing the most specific characteristics of the *desert state*, as is found, to varying degrees, in all deserts around the world.

The characteristic that can be spotted by even the untrained eye is the *rapid degradation of any green cover*. If someone were flying over it, in just a few minutes the view of greenery from the plane would shift from a clear forest or savanna to a disjointed steppe until it disappears altogether. On land, however, we can see that this degradation involves a gradual decrease in both the number of individual plants as well as the number of species (biodiversity), leading to increasing homogeneity. This is compounded by the unequal distribution of vegetation, spread out on the slopes and interfluves, concentrated in the *wadis* and showing significant variations in vitality depending on the rainfall, with certain plants rapidly blossoming after a downpour. This decrease in vegetation is accompanied by an equal decrease in the fauna in the region, especially of the animals that are most dependent on vegetation and water.

The *hydrographic disruption* manifests as the rarefaction of drains and a growing anarchy in water flows. Concentrated drainage divides into independent networks that become fewer and fewer in number and less and less hierarchic, with the water finally forming isolated channels or even disappearing completely. Most channels are dry in all seasons and are only revived during a flood. Apart from the large, exorheic[4] allogenous rivers, such as the Nile or the Euphrates, water flow is exclusively local and linked to precipitation, which is itself intermittent and irregular. Between channels, during the rains the surface of an interfluve may often be swept by diffuse flows, in sheets or numerous and changing gullies. All these drains, except for those along the maritime coasts, are endorheic[5], that is, they lead to closed basins of all sizes, which are numerous and independent, where they are lost. There is also the extreme case of an arheic region[6] where there is a complete absence of water flow.

4 The term "exorheic" describes a river that drains into the sea or another water body.
5 The term "endorheic" signifies that a body of water terminates in a closed depression.
6 The term "arheic" describes a temporary body of water with no surface runoff at all.

The morphology of a desert is also formed of unique features of relief, owing to the specific conditions of the desert model. The disaggregation of hydrographic networks reduces the preeminence of valleys in a landscape. The desert is flat until it enters the mountains. The horizontal shapes or shapes with very small slopes, plains, plateaus and glacis[7] overpower sharp reliefs or reduce them to eroded tussocks or to small isolated massifs (*inselbergs*[8]). And everywhere you have evidence of the wind at work. The sweeping winds shave off the finest elements from the rocky floors and alluvial sheets and accumulate them into dunes, scattered or grouped together (*ergs*).

The scarcity of water and arable land leads to a *dispersion of land use and sedentary life*, with sedentary populations tending to settle around better land, or land that is less poor. Humans seek alluvial land, which is more or less regularly flooded or is, at least, irrigable using ancient and varied systems. This leads to populations, cultures and habitats (*oasis*) being scattered (around water bodies) or arranged in a linear manner (along a river), further accentuating the powerful impression of a void that a desert creates. A relative void, however, because although the isolated and precarious water bodies, springs or wells cannot sustain agriculture by themselves, they can sustain an active, if highly scattered, nomadic life.

To summarize, we can identify five factors that make up the concept of a *desert* and that characterize deserts, to varying degrees and based on various combinations of different geographic features:

1) the most general of these is *aridity* (see Chapter 3, p. 65), an obvious aspect that is, however, difficult to define and measure, as it involves the scarcity of precipitation, intense evaporation as well as the nature of soil (substratum and superficial formations);

2) problems with *water* (see Volume 2, Chapters 1–3) are largely dependent on the previous point. While these problems involve the modalities of flow and how these affect the landscape morphology, they also involve water inputs and water capacity in the region, as well as the various ways in which inhabitants of the region use these resources;

3) the presence of *salts* (see Volume 2, Chapter 4), whose nature and abundance depends on the highly varied combinations of aridity and the input of dissolved salts in water, a condition that makes the water non-potable but also serves as an essential resource. This mineral wealth can be transported by the caravans for trade;

7 A "glacis" is a plane or slightly concave surface, with a small slope that is less than 10°, at the foot of a steep slope.

8 An "inselberg" is an isolated hill or mountain rising abruptly from an erosion plane.

4) the *wind* (see Volume 3, Chapters 1 and 2) is omnipresent and especially strong in the desert. All along its complex paths it lifts up, transports and deposits sands and dust;

5) and finally, an essential factor is that of *space*, which contains all the other factors. There is, first of all, the large expanse, an inordinately vast area in some deserts. Then there is also the bareness of the biological environment. The relief forms, which depend on unique legacies which, to greater and lesser extents, are ancient. Finally, you have the inhabitants, both sedentary and nomadic, who have, over the centuries, been able to construct and maintain a veritable "desert civilization" (see Volume 3, Chapter 3).

Over the course of these volumes, we will examine these five basic factors in detail. However, we begin with an overview of the history of our exploration of the deserts.

1.4. Deserts of the world

If it is not easy to define a *desert*, it is even harder to establish its boundaries. Most often, we see a progressive transition from the humid to the arid and then onto the desertic. These nuances are themselves limited by fuzzy boundaries, with capes, islands and enclaves that change with seasons and the passing of the years. We can ascribe certain properties to them that are more or less measurable: climatic (the abundance of rain, indicators of aridity), hydrological (density of drainage, flow indices, endorheism, arheism), biological (plant cover, composition of the flora, adaptation to dry conditions, endemism[9]), morphological (relief, geodynamic systems) or even economic (presence or absence of cultivation, the need to use irrigation). Each of these arguments or criteria is both valid and disputable, but they are only truly meaningful when combined with the others in an overall network of interdependencies that make up the desert ecosystem.

Thus, citing a specific figure for the area covered by deserts around the world is both an illusory and approximate exercise. This being said, we can broadly say that deserts (in the broadest sense) cover about a third of the area made up by the continents, without taking into account frozen deserts. They are distributed across the globe in inter-tropical and temperate regions, along three basic axes (Figure 1.1):

1) the Afro-Asiatic Eremic diagonal[10] that includes the Sahara and the deserts in Asia;

---

9 Endemism is used to describe species that are unique to a particular region.

10 The term "eremic" means pertaining to deserts, from the Greek ἔρημος (érèmos) = desert, solitary.

2) the North and South American Eremic diagonals;

3) the isolated deserts in the Southern Hemisphere: Northeast Brazil, South Africa, Madagascar and Australia.

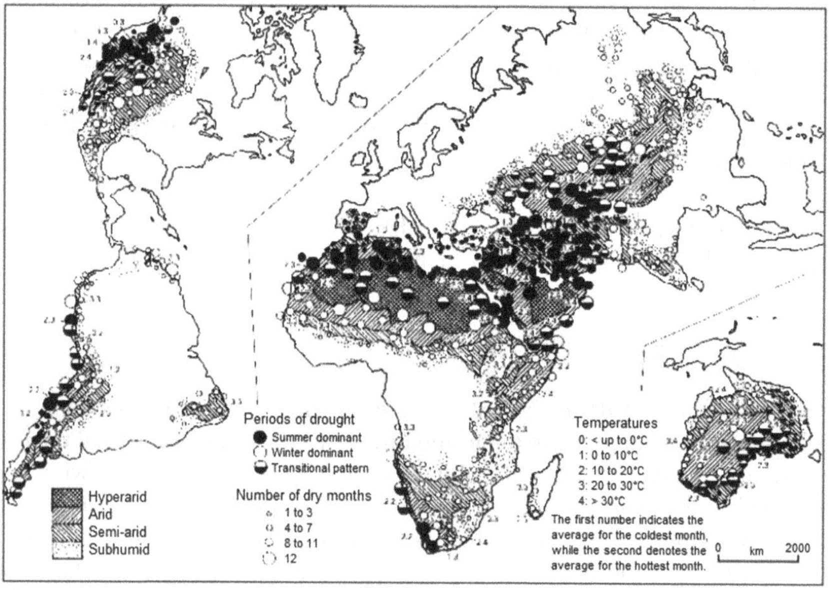

Figure 1.1. *Arid regions around the world as per (Meigs 1977–1979), modified*

How can it be that, in the age of air travel, widespread tourism, television and the Internet, we still so often have such a fuzzy, conventional and mythical image of the deserts? There has certainly been progress in gaining direct knowledge of the deserts, but this knowledge itself has spread very slowly and imperfectly to non-desert regions. The difficulties around entering and traveling through the desert can be blamed for this slow dissemination, as well as the risks inherent to these regions. Man has traveled across the deserts for centuries out of necessity and as a result of certain events (military, political or economic), rather than through a desire to discover or gain knowledge. The merchants and travelers would follow linear paths marked by water points. They would rarely deviate from these paths, for fear of losing themselves and dying in the desert. Modern automobiles (not counting those used in the rallies) can cover in one hour the distance that a camel would take a day to travel. Those who make these journeys today do so at a speed that allows them to grasp only the most superficial of information. Views from planes or satellite images are always difficult to decipher. Tourists, reporters and artists, who are very demanding, prefer the picturesque or sensational over banal reality. There is an increasing number of scientists in this field, but they have a very limited audience. For all of these reasons, the image that the public receives of the desert is partial and varies widely depending on the source. The varying levels of reliability of the information and the unequal distribution of this knowledge go a great way toward explaining the obscurity, mystery and legends that deserts are veiled in even today.

1.5. To know more

Birot, P. (1970). *Les régions naturelles du globe*. Paris: Masson.

Chenu, R. (1997). *Le désert — Petite anthologie*. Paris: CERF, 129 p.

CRISM, ed. (2000). *Figures — Le désert*. Cahiers du Centre de Recherches sur l'Image, le Symbole et le Mythe, Université de Dijon, 167 p.

Cuny, H. (1961). *Les déserts dans le monde*. Paris: Payot.

Demangeot, J. and Bernus E. (2001). *Les milieux désertiques*. Paris: Armand Colin.

Doucey, B., ed. (2006). *Le livre des déserts*. Collection Bouquins. Paris: Robert Laffont, 1231 p.

Dresch, J. (1966). "La zone aride, Géographie générale." In Encyclopédie de la Pléiade. Paris: Gallimard, pp. 712–780.

— (1982). *Géographie des régions arides*. Collection Le Géographe. Paris: Presses Universitaires de France. 277 pp.

Haudricourt, A. and Hédin L. (1943). *L'homme et les plantes cultivées*. Géographie humaine. Paris: Gallimard. 233 pp.

Hills, E.S., ed. (1966). *Arid lands, a geographical appraisal*. Paris, London, UNESCO, Methuen & Co. 461 pp.

Joly, F. (1957). "Les milieux arides— Définition— Extension." In *Notes marocaines, Rabat* 8, pp. 15–30.

Mc Ginnies, W., Goldman, B. and Paylore P. (1968). *Deserts of the world — An appraisal of research into their physical and biological environments*. University of Arizona Press. 788 pp.

Meigs, P., ed. (1977–1979). *Carte de la répartition mondiale des régions arides, Carte à 1/25 000 000, 1 feuille en couleurs, Notice explicative*. Notes techniques Man and Biosphere 7. Paris: UNESCO. 55 pp.

Monod, T. (1937). *Méharées. Explorations au vrai Sahara*. Bibliothèque des voyages. Paris: Je sers. 300 pp.

— (1973). *Les déserts*. Paris: Horizons de France.

Petit-Maire, N. (1984). "Le Sahara, de la steppe au désert." In *La Recherche* 160, pp. 1372–1382.

Petrov, M. (1976). *Deserts of the world,* New York: John Wiley & Sons Ltd.

de Planhol, X. and Rognon, P. (1970). *Les zones tropicales arides et subtropicales*. Paris: Armand Colin. 487 pp.

Pouquet, J. (1951). *Les déserts*. Que-sais-je ? 500. Paris: Presses Universitaires de France.

Sorre, M. (1943). *Les fondements biologiques de la géographie humaine — Essai d'une écologie de l'homme*. Paris: Armand Colin.
UNESCO, ed. (1962). *Les problèmes de la zone aride — Actes du colloque de Paris*.
— ed. (1995). *Les zones arides dans les programmes de l'UNESCO*. Paris: UNESCO.

2

Conquering Deserts

"Force this space and solitude, for it will be worth the sacrifices made."
Antoine de Saint-Exupéry

2.1. Prehistoric times

Our earliest contact with the arid world of the desert can be traced back to the very origins of humanity. It is generally accepted that humans originated in East Africa, in the African part of the rift valley called the Great Rift Valley[1]; somewhere between Tanzania and Ethiopia around Lake Turkana (also known as Lake Rudolf) and Lake Victoria. This is a region that has always abutted arid spaces.

In the second half of the Cenozoic era (third age), during the Miocene and Pliocene epochs, the slow drift of the African lithospheric plate toward the northeast, caused by the expansion of the Atlantic ocean, ultimately ended up positioning the Great Lakes region such that they straddled the equator, between the latitudes 10°N and S. Across East Africa, this movement was accompanied by increasing dryness. The dense tropical forest, which still exists in equatorial West Africa, here gave way to a dry savannah, dotted with trees or grassy steppes with spiny bushes, which are also still seen today.

Chapter written by Fernand JOLY.

1 The Great Rift Valley: a series of contiguous tectonic trenches running from north to south from the Mozambique canal all the way to the Red Sea, extending into the trough containing the Dead Sea, with intense seismic and volcanic activity corresponding to the opening of an ocean, in what is today the Red Sea.

It was in this arid or semi-arid bioclimatic environment, 15 or 20 Ma[2], that a distinct species emerged among primates, independent of monkeys, widely considered to be the precursors of the human race. Between 5 and 1 Ma, this species developed into a non-arboreal bipedal population that still presented certain simian traits. First discovered in southern Africa, these prehumans were given the name "Monkeys of the South" or *Australopithecus*. They were wandering hunters, evolving in an open landscape amidst large herbivores (elephants, giraffes, rhinoceros, antelope, etc.) and sometimes using tools made of rough stone or flint flakes, gathered from the soil.

In the latter part of the Pliocene epoch, between 3 and 1.8 Ma, in East Africa again (Ethiopia, Kenya, Tanzania, South Africa), in an arid but non-desert environment there appeared, jointly with the *Australopithecus*, the first species which we qualify as human. Prehistorians created the genus *Homo* for this line. These were again nomadic hunters, but armed with pebble tools that were deliberately chosen (*pebble culture*) and, thus, they were called *Homo habilis*. It is also suspected that this species lived in occasional or temporary shelters and may even have had a nascent social structure.

The climatic changes multiplied at the end of the Pliocene and the early Fourth Age (lower to middle Pleistocene) between 2 and 0.2 Ma. These changes marked the start of the first Ice Age, which manifested in Europe and North America by alternating cold periods (glacial) and warmer periods (interglacial); and on the arid edges of Africa as relatively humid periods (pluvial) and drier periods (interpluvial) with an expansion of the deserts. About 1.8 Ma, in the open savannahs in East Africa, where the climate was tempered by the first quaternary rainstorms, there appeared a new population of real *humans*, initially incorrectly called "Monkey-Men", or *Pithecanthropus*, and then more correctly identified as *Homo erectus*, because of their ability to stand upright and walk.

Homo erectus, which had a larger brain than its predecessors (around 950 cm^3), was able to use tools cut from stone, fashioned in various ways (biface with sharp edges for cutting, tools for scraping, pointed tools, etc.) and around 0.5 Ma, they started using fire. Above all, these hunters spread out widely across the peri-desertic steppes beyond East Africa seeking out their prey: herds of herbivores (antelope, cattle, horses, etc.). Bifaces have been discovered almost everywhere in the Sahara, abandoned on the ground. Relicts of *H. erectus* have been found in North Africa, as far as the Atlantic coast. Skirting the deserts of Arabia, Syria and Iran, *H. erectus* then entered Europe and Asia, reaching China, Southeast Asia and Java.

2 Ma: million years ago.

The "humanization", in both biological as well as cultural terms, accelerated during the upper Pleistocene period between 200 and 10 ka BP[3]. Toward 120 ka, during the long Riss-Würm interglacial period – which was divided into a series of several pluvials and interpluvials, sometimes nearly wiping out the deserts – *H. erectus* disappeared and in its place emerged "modern man", i.e. *Homo sapiens*. This was a species that was more physically evolved, socially organized and intellectually astute (diversified tool-making, using stone and bone, the development of abstract thought and art, funerary rites, language, etc.). The oldest of these new beings was *Homo neandertalensis*, who flourished in Africa, the Near East and Europe between 80 and 40 ka, at the start of the last glacial period (Würm). This species, in turn, disappeared (or evolved) between 35 and 30 ka, giving way to their contemporaries and then successors, *Homo sapiens*, the ancestors of modern humans.

At the end of the Würm glaciation (40–10 ka) – while humans in Europe were subject to harsh climates to which they adapted with a great deal of ingeniousness – groups of hunters continued to travel around the semi-arid steppes, located between the periglacial tundra and the subtropical deserts. Through Palestine, Syria and Iran, they reached Turkestan, India and Southeast Asia; they can be traced as far as Australia. Other groups circumvented the Himalayas and the Chinese deserts, entering East Asia (Siberia, Mongolia, Kamchatka). From here, between 50 and 30 ka, they went into North America through the Bering isthmus or along the path created by the Aleutian islands, which opened up as water retreated and froze into ice.

Using largely arid, peri-desertic spaces, Homo sapiens thus pursued the conquest of the whole of planet Earth.

2.2. The dawn of history

The postglacial period (Holocene) began about 10,000 or 12,000 years ago when the ice sheets covering northern Europe, Siberia, Scotland, Iceland, Greenland and North America retreated. The ice gradually melted as the climate grew noticeably warmer, with an increase in precipitation in the arid regions in the Northern Hemisphere and a general rise in sea level (the Bering isthmus became the Bering strait).

The drought that prevailed in the Sahara at the end of the Pleistocene period receded more rapidly in East Africa and the Near East than in the western Saharan region. The climatic optimum (moderate temperatures and adequate rain) set in

3 ka: thousand years ago; BP: before present, by convention, taken to be the number of years before January 1, 1950.

between 9,000 and 5,000 BCE. From this time onwards, the Afro-Asiatic deserts grew steadily drier, despite some occurrences of brief and low-intensity rainfall. On the other hand, many important technical and cultural innovations transformed human life around the periphery of these deserts, or in their mountains, and all along the large rivers that flow through them (the Nile, Euphrates and Tigris).

Until this point, humans had survived almost exclusively by hunting and gathering and sometimes through fishing along rivers, lakes or seas. The first attempts by humans to domesticate wild animals were made in the Holocene epoch, in the course of their wandering life in search of herds of herbivores in the open, peri-desertic regions. First dogs, present since the end of the Ice Age, and then smaller livestock like the goat and sheep, followed by cows, pigs and then the ass, horse and camel (dromedary). All were descended from the fauna of the steppes or dry savannahs.

This period also saw the earliest attempts at domesticating plants, first through the selection of useful species from among those that grew wild over the steppes, and then through agriculture. This new relationship between humans and the earth was accompanied by a gradual tendency among some of the nomads to settle down around arable land, either close to water sources or in the flood plains. The sedentarization and the consequent improvement in the standard of living was marked by a notable increase in population, a rudimentary farming life and the emergence of the first villages.

If we look at the inhabited land in the West, it is again along the borders of the deserts, if not within the deserts themselves, that we see the first cradles of civilization.

2.2.1. *The Near and Middle East – Sahara*

It all began in the Middle East between 9,000 and 5,000 BCE. During this period, it was possible to harvest wild grains, barley or wheat on the semi-arid slopes (200–400 mm of rain) of the western Zagros and central Taurus mountains. The first farmers here were able to use spring water and water courses that began in the mountains. Very soon, villages were established in what would later be called the “Fertile Crescent”, around the Syrian desert. From this time onwards, during the climatic optimum that set in the Holocene from 7,000 to 6,000 BCE, agriculture spread all throughout the floodplains around the Tigris and the Euphrates (Mesopotamia) and in the Jordan valley, thus creating the first oases. Humans learnt to cultivate cereals: the earliest types of wheat (emmer wheat, durum wheat and common wheat) as well as barley, millets and oats, grown in the shade of the date palms that grew naturally in this region.

Despite the dry climate, the period between 4,000 and 3,000 BCE represents a key period of development in the arid Middle East and Near East. The invention of irrigation and the use of draught animals (bulls and donkeys) allowed the farmers to cultivate more land and to diversify the crops grown. In addition to the cereals, they now grew legumes (beans, peas and lentils), textiles (flax and cotton) and fruit trees (fig, pomegranate). The rise of subsistence farming and the arrival of new immigrants (Semites from the Arabian desert, Sumerians from Iran or Armenia) led to a considerable rise in the number and concentration of "brick" villages. Some of these villages grew until they became cities, such as Ur, Uruk and Eridu, which were located at the mouth of the Euphrates; or Jericho, in the Jordan valley. The deposits of silt and clay in the alluvial plains and the marshy reed beds provided enough raw material to make bricks and adobe, ceramics and pottery. However, they lacked stones for construction, cloth for dressmaking and ores that could have been used for early metallurgy. Necessary exchanges were established with their neighbors: while Egypt and Iran were chief among these, there was also Anatolia, Phoenicia, on the coast, and through Phoenicia, the Mediterranean region. Toward the start of the 3rd millennium BCE, arid Mesopotamia was the chief economic and cultural center in the Western world.

The rise of Egypt and Nile Valley Civilizations can be compared to Mesopotamia. Just as Mesopotamia was "a gift of the Tigris and the Euphrates", Egyptian civilization was the "gift of the Nile". A long river, a large alluvial plain, and a marshy delta counteracted the harsh desert conditions. The drying of the climate, which began earlier and was more continuous than in Asia, was not disrupted by a small, insignificant climatic optimum. These conditions only made the valley more attractive to the Saharan nomads and sedentarization started as early as 8,000 BCE. A population of farmers and fishermen gathered on the Nile terraces, in the humid oases in the Libyan desert (Kharga, Dakhleh, Fayoum around Lake Moeris, Wadi Natroun) and on the alluvial deposits on the Delta. From 8,000 to 3,000 BCE, Egyptian farmers used the rich range of locally available species (spelt, durum wheat, barley, millets, flax; small livestock, pigs, cattle, zebu, ass) to which they added exotic species from other regions: sorghum and cotton from Nubia or Sudan, date palms from Mesopotamia, common wheat and legumes from Iran, sesame from Abyssinia or India and, later, the horse from Asia, the Arabian dromedary, olive trees and vines from Greece and so on. Settling in villages built up on adobe or stone, and located away from the floods, Egyptian agriculture flourished at the end of the Neolithic era because of the regular flooding of the Nile.

The case of the central and western Saharan regions is more surprising, as these regions also witnessed the growth of a remarkable rural civilization. This part of the Sahara, which is today rightly considered the model for a desert, had a pluvial climate for much longer than Libya or Egypt – until the middle of the 3rd millennium BCE – and consequently almost the entire region was habitable. Pottery dating back over 9,000 years has been found in the Hoggar mountains and it has been established that from 8,000 BCE onwards there was widespread agro-pastoral activity in the central Saharan region. Traces of habitation found in caves, the remains of huts, of lakeside villages and rock engravings in caves all reveal the existence of a population of hunters, breeders, farmers and fishermen. This civilization seems to have reached its peak around 4,000 BCE. Originating, undoubtedly, in the Middle East and having traveled through Cyrenaica, Fezzan and Egypt, this population lived notably in the Hoggar, Tassilis, Tibesti and Ennedi regions, where small hydraulic structures have been identified, implying the existence of subsistence farming. From here, it grew toward Mauritania and Mali, on one side, and toward Aïr, Bornou and Chad on the other. However, from 3,000 BCE onwards, increasing dryness gradually began to confine agricultural life to oases in the desert.

Thus, between the 9th and 3rd millennium BCE, the arid space between the Sahara and Iran saw the first significant technical, economical and cultural revolution in human history. In fact, this could be termed an evolution rather than a revolution, since it took place in a more or less continuous manner over the course of 50 centuries. The fact that this took place in an arid region cannot be by chance alone – the phenomenon brought into play a number of physical and human factors that were conducive to common development across these regions, although aspects of it were determined by local conditions.

– We cannot discount the role played by climate. The progressive increase in dryness, starting from a state that was already close to aridity, did not produce changes that were as drastic as those produced by the melting of glaciers in Europe. The retreat of green cover was slow and the steppes, the favored path taken by the nomadic hunters, also provided the first animals used for rearing and agriculture. Indeed, the basic animals and plants that are used even today were already in place at this point. In any case, a fundamental consequence of the increasing dryness was that it encouraged the sedentary populations to concentrate around the oases, formed on the richest, irrigable land. This led to the development of a rural village society (which later became urban society), which was both cooperative and imaginative.

– Stone tools, which developed through the prehistoric period, were perfected and diversified (polished stone, specialized, miniaturized tools) and persisted until the 3rd millennium BCE. We find specimens buried in the alluvium in Lower Mesopotamia, the Nile Delta, in the villages along the Fertile Crescent and along the edge of the Libyan plateau. This "Neolithic culture" was enriched over time by the use of other materials: bone, horns, ivory, pottery, ceramics, leather and metals. Egypt is likely to have had knowledge of gold, silver and copper from the 8th millennium BCE onwards, while bronze appeared in Mesopotamia and Egypt in the 3rd millennium BCE. From the 5th millennium BCE onwards, metallurgy and irrigated agriculture were the two advanced techniques that characterized protohistoric civilizations in the East.

– This evolution had significant consequences. Historians have emphasized, with reason, that this represents the passage from a hunter-gatherer economy to a production economy and this transition became the norm. Movement of peoples and cultural and trade exchanges intensified between Mesopotamia, Egypt, Anatolia, Iran and even India. Routes opened up toward almost every destination through the Afro-Asiatic deserts. Intellectual and artistic life developed and grew, chiefly in Egypt and in Mesopotamia.

– The 3rd millennium BCE marked a turning point in history. Political and military conflicts between urban communities or kingdoms, to establish a hegemony, finally ended (around 3,000 BCE) with territories being divided and the foundation of the first dynasties in Sumer, in Lower Mesopotamia, and Memphis, in Egypt. The invention of writing (also around 3,000 BCE), in the form of hieroglyphics in Egypt and cuneiform in Babylon, would facilitate both the transmission as well as the preservation of knowledge and thinking, and this development is conventionally taken to mark the start of the historic period.

There is generally less information on the prehistory and protohistory of other deserts in the world, either because the study of these deserts began later, or because there were fewer protagonists and fewer documents came to us. What information we do have, however, still reveals certain characteristics that are common to all arid regions and that are worth highlighting.

2.2.2. *Central and East Asia*

The deserts that extend from the Caucasus range to China are made up of arid or, frankly, desertic basins, surrounded by high mountains. Long glacis of piedmont that stretch between high land that is hard to access or steeply cut away and the dry, bare or sandy plains, are covered everywhere, by a grassy steppe that has always been conducive to grazing, pastoralism and migrations. The numerous Paleolithic and Neolithic sites here reveal the existence of nomadic pastoral communities that can be traced back to the end of the Ice Age and that persist even today.

The cold continental steppes, with hot summers and freezing winters, extending from north of the Caspian Sea all the way up to the Mongolian steppes, may not have produced edible crops but they provided the Indo-European (Aryan) and Paleo-Asiatic (Mongols) hunters and livestock breeders with the ideal animal for covering the large, flat regions: the horse. This animal could be used harnessed to a vehicle (the chariot on wheels is likely to have originated in the North Caucasus region toward the 4th millennium BCE) or used as a mount. The shepherds in the warm or temperate steppes in Iran or Chinese Turkestan domesticated camels and dromedaries for use as pack animals. Those who lived in the mountains and high plateaus tamed yaks. These communities reared herds of sheep and goats or cattle along the edges of large deserts: the Lut, Karakum Kyzylkum and Taklamakan.

Agriculture did not really emerge here before the 3rd millennium BCE as neither the climate nor the soil conditions were favorable. While Central Asia had few contributions to make to neighboring agro-pastoral civilizations, apart from draught animals and pack animals, it received many things from these communities: cereals from the Middle-East, wheat, sesame and hemp from India, soya and vegetables from China, and irrigation techniques from Mesopotamia and Egypt. Agriculture here, however, remained scattered across isolated oases in valleys and mountains, and around water sources in the alluvial plains. Among these, certain regions saw much greater development than others as a result of more advantageous local circumstances. This was the case in the warm steppes of eastern Iran, which were sensitive to the influence of Mesopotamia. Grain silos and remnants of cultivated fields have been found around the Qom desert (the Great Salt Desert or Dasht-e-Khavir) from the 5th millennium BCE onwards, and to the east of the Lut desert dating back to the 3rd millennium BCE. This was also the case in the Indus Valley in present-day Pakistan, along the border of the Thar desert, which was inhabited from the 5th millennium BCE onwards and was the center of an advanced urban civilization which lasted until the end of the 3rd millennium BCE. While not as advanced as this, the cold eastern steppes also had flourishing agro-pastoral civilizations with influences from China and India. Oases emerged in valleys to the south of the Tian Shan and in the Tarim basin. Some of these are sites that are occupied even today. The Loess plateau[4] in the Ordos loop of the Yellow River was the cradle of a Palaeolithic civilization that existed around 30,000 years BP and in the Neolithic period (from the 4th millennium BCE onwards), it was the seat of a prosperous agricultural society, and a thriving cultural and political center, to rival the earliest Chinese dynasties.

Here, as in the West, increasing dryness in the climate accelerated sedentarization and even urbanization. From the 2nd millennium BCE onwards, there had been a steady decline in pastoral nomadism in the face of an ever-increasing concentration of populations around certain sites. Large-scale migrations took place toward more favorable regions in the Middle East, Europe and North India. Certain oases were abandoned – this may have been due to dryness but it is more probable that it was due to pressure from the "people of the Steppes" (who later became the Mongols and the Turks). These confirmed nomads (who traveled the open lands between the Aral Sea, Lake Baikal, Dzungaria and the Gobi Desert) were fierce warriors. Their raids along

4 Loess: referring to the "loose" yellow soil deposited through windstorms during the cold and dry phases.

the trade routes and their invasions of newly emerging States around them would shape the history of the continent until at least the 16th Century CE.

Indeed, once the initial civilizations were set up, the conflict between empires would be the chief means of conquering the Afro-Asiatic deserts.

2.2.3. *America*

Turning now to the two American continents, little is known about the relationship between humans and the American deserts before the Spanish conquest of the region. The absence of texts and the limited use that the Conquistadors made of oral tradition (wherever they thought it of interest) make it difficult to get a clear picture of how things stood at the dawn of the historical period here (the late 15th Century CE). Consequently, there is imperfect and uncertain knowledge about the arid zones of America in the pre-Columbian era. What knowledge there is has been gleaned chiefly through geological, paleontological and archeological discoveries, which have been rare and widely dispersed.

Based on their anatomical or morphological characteristics, the American "Indians" or Amerindians are likely to have come from Asia (Siberia or Kamchatka) between 50,000 and 30,000 years ago. With the melting of the North American ice sheet, they most likely then spread out eastwards, toward the Great Lakes and the Mississippi basin, and westward through the steppes and deserts between the coastal ranges and the Rocky mountains, toward the high plains of Utah, Nevada, Arizona, California and New Mexico. The oldest remnants discovered in this region are houses and stone tools associated with mammoths and bison that date back at least 15,000 years ago. Moving southwards, this population is likely to have reached Mexico around 14,000–12,000 years ago and the southern tip of South America (Patagonia) around 10,000 years ago. From this point onwards, over the millennia, the groups diversified both in terms of facial features, as well as their language, lifestyle and social and political organization. A question that remains open even today is whether this diversity resulted from an evolution *in situ* or through successive waves of invasions from Asia and/or Oceania.

When the Europeans arrived in the Americas, the territories between the two tropics on either side of the equator were divided between three great empires: the Mayan Empire in Yucatán and the American isthmus, the Aztec Empire in Mexico, and the Incan Empire in the Andes in Peru. These empires mostly occupied the mountainous, humid and forested regions. The desert, on the other hand, stretched across a large part of the Aztec empire, from the cacti-filled steppes in the north to the Chihuahuan desert in the central Mexican marshlands, as well as over the Peruvian Andes and into Bolivia, to the south of the Incan Empire. However, unlike the Afro-Asian civilizations, which were born in the deserts or on the edges of deserts, in the Americas civilizations first emerged in the humid, tropical regions and then spilt over into the neighboring arid zones.

While the Aztecs, who came in from the north, remained largely nomadic hunters, they introduced irrigated maize farming in the southern part of their kingdom, under the influence of the Mayan Empire. The Incas practiced terrace farming on alluvial fans[5] and the large, dry slopes of the Andes, from Guyaquil to Chile. They also colonized and controlled the Bolivian Altiplano delta and the Chilean Atacama, more or less ruling over the nomadic hunters of Chaco and Patagonia.

In arid North America, on the contrary, most Amerindians remained outside the great Central American empires. In the southwest of present-day United States (New Mexico and Arizona), sedentary farmers lived on the cliffs of the canyons or in large stone and brick villages, now known as pueblos. They practiced pluvial or irrigated cultivation of maize, vegetables and cotton, representing the northernmost advance of Aztec and Mayan techniques. Nomadic hunter-gatherers traveled through the arid basins to the west of the Rocky Mountains. Some of them practiced fishing in the lakes or along the coast. The populations in the vast, semi-arid plains between the Mississippi and the Rocky Mountains, who hunted mostly buffalo, maintained some oases along the valleys.

Although the domestication of plants by the Amerindians could be compared to that carried out by Neolithic civilizations of the Ancient world, in reality, in the case of the Amerindians, the arid zones had very little role to play in the process. The primary crops, such as maize (*Zea mays*), manioc, sweet potato, cacao, beans, sunflower and cotton, originated in the tropical lands. Some of these crops are likely to have spread through the Incas and the Aztecs as far as the oases in the deserts. The driest regions, on the other hand, provided crops like the Andean potato, melons, agaves and opuntia. Tobacco was reasonably widespread. The only domesticated animal species were the dog, the barnyard turkey and, on the Peruvian-Bolivian Altiplano, the llama (as a pack animal) and the alpaca (for wool). The Amerindians had no knowledge of either the horse or the wheel. They produced ceramics and pottery, and made items out of bone, horn, leather and textiles; however, their knowledge of metallurgy was limited to copper, bronze and gold, and their tool-making never went beyond Stone Age tools. These empires achieved a very high degree of political, artistic and intellectual culture, however their technological prowess was no match for the arms available to the conquistadors in the late 15th Century. In general, outside the empires, civilizations in the arid regions of the Americas remained very basic. While it is true that hunter-gatherers dominated the desert spaces, they were strongly divided into warring tribes whose relationships with

5 Alluvial fans: alluvial deposits formed by a river whose slope decreases sharply as it flows out into a valley.

each other were complex and often driven by conflict. The Europeans first came into contact with the Empires. The conquest of the deserts would only take place decisively with the establishment of the neo-American states in the 18th and 19th Centuries.

2.2.4. *Southern Africa – Australia*

The deserts of southern Africa and Australia are among those that were studied only toward the 18th and 19th Centuries, with the expansion of European colonization. These deserts also hold only paleontological and archeological records. How prehistoric humans conquered these lands and settled here remains relatively unknown. However, available information is enough to reveal common characteristics between these territories, as well as significant differences, at least over the last hundred millennia.

Southern Africa and Australia are both continents in the Southern Hemisphere, 81% of which is oceanic. It also includes two deserts located less than 45° of latitude from the huge Antarctic ice sheets that are centered around the South Pole and are over 15 million years old. This created certain conditions that were common to the climatic evolution of both of these deserts, specific to them and distinct from those of deserts in the Northern Hemisphere.

Despite these commonalities, however, there are also significant differences between these two territories in terms of biological development and inhabitation by human populations, both of which were influenced by the geographical conditions unique to each of these deserts. After the breaking up of the continent of Gondwana (toward the end of the Jurassic period and the early Cretaceous period, 170 Ma), southern Africa remained attached to the African continent, drifting northeast. Hominids, and then humans, were therefore able to travel freely across this region. On the other hand, although Australia had been attached to New Guinea and to Tasmania throughout the last ice age (70,000–30,000 years ago), it has always been isolated with respect to neighboring Indonesia and the continent of Asia. Furthermore, although this isolation led to a unique paleontological evolution of Australian flora and fauna, this evolution did not go as far as the human species. Prehistoric humans in Australia are thus likely to have come here by sea, perhaps 40,000 to 30,000 BP.

When Europeans settled in the Cape region of southern Africa, in 1652, the deserts located further north were practically unknown to them. It was only in the 20th Century (1924) that the first *Australopithecus* skull (about 3 or 4 million years old) was found in a pit in Bechuanaland in Taung, in the southern part of the Kalahari desert. Since then, many other remains have been found along the southern and southeast borders of the desert. These include remains of *Australopithecus*,

H. erectus as well as *Homo sapiens*. It can thus be seen that the history of human populations in the deserts of southern Africa go back deep into prehistoric times, and that this population lived quite obscurely, producing no notable innovations. In the 17th Century, the arid territories extending between 15 and 30°S (from Angola to the Karoo) and along the Atlantic coast (Namibia) to 30°E (the Veld plateau) were largely inhabited by nomadic hunter-gatherers, with conflict arising between tribes: the Bushmen toward the center (Kalahari) and the west (Namibia) and the Khoikhoi (Nama) in the south and along the coast. However, for a long time these indigenous tribes had been progressively pushed back by the Bantu tribes from the north east (Zimbabwe) and the north (Angola and Katanga), who were propagators of agriculture, cattle breeding and the metallurgy of iron. The Bantu peoples mainly occupied the semi-arid fringe of the deserts from the Veld (Zulus) up to southern Namibia (Hereros). The Kalahari basin in the center was traversed by Bushmen and remains uninhabited and completely uncultivated.

The oldest Australians are the *Homo sapiens* who came from the north through Java and New Guinea. These humans must have been among the earliest navigators and they settled along the coast up to Tasmania, probably even along those regions on the continental shelf that are submerged today. They are likely to have practiced fishing and harvesting shellfish and moved inward for the humid climate by going up rivers and identifying lakes. They thus arrived at and settled in the Murray-Darling basin and the depression of Lake Eyre. This region, which is arid today, is the richest site in Australia in archeological remains from the Pleistocene (beyond 10,000 BP). These remains reveal bone fragments and stone tools (cores and flakes). The rise in sea levels during the Holocene accentuated the earlier move to shift inland, toward the center (the Simpson desert) and westwards (the Victoria desert), where sites have been found with retouched stone tools. These earliest occupants, the ancestors of present-day Aboriginals, found neither cultivable plants nor animals that could be domesticated. They must certainly have imported dogs, some of which returned to the wild (the dingo). Even in contemporary times, the Aboriginals, who are neither herders nor farmers, and armed only with spears and boomerangs, remain nomadic gatherers, hunting emus and kangaroos.

2.3. Knowledge of deserts in prehistory

If the historic period is the period in human evolution that we know about through written or carved texts, then the prehistory of deserts can extend to various times, sometimes even reaching up to the 19th Century. During this time, these arid zones were conquered by hunter-gatherers and then nomadic pastoralists. As agriculture emerged, one group of nomads settled in privileged areas such as oases or villages, which were geographically isolated but connected through complex networks of traffic and influence. This sedentarization stimulated local growth in cultural centers, cradles of civilization that were often very advanced and technologically innovative.

Thus, over thousands, if not millions of years prior to their entry into recorded history, humans undoubtedly learnt a lot about deserts and had acquired considerable experience of the arid world. It is difficult to say how the inhabitants of these regions were benefited from their experience and how far it may have traveled into non-desertic regions, since we do not have any direct record. As they were probably aware that they belonged to the same environment, perhaps these people were more sensitive to diversity in language, beliefs and political situations than to the perception of geographic nuances? The deserts of this remote era have left us only scattered paleontological and archeological data: human bones, several stone tools that have sometimes simply been abandoned on the ground, the remains of houses and camps, foundations of buildings, early works of art (jewelery, ceramics and statuettes) and also the marvelous "Library of Stone" that Monod describes, composed of rock engravings that depict the fauna and daily lives of the humans of this period. Finally, the most common evidence of prehistoric deserts is geological and geomorphological in nature: traces of ancient hydrographic networks, sedimentary formations of fluvial, riparian, or wind-deposited matter and the inherited landforms.

2.4. Antiquity

2.4.1. *The Near East – Sahara*

In the Near East, in the 3rd millennium BCE, the Mesopotamian basin was occupied by indigenous Semitic farmers. In the south, these populations mixed with immigrants who were undoubtedly from the neighboring mountains (the Sumerians), and in the north, there were Semitic invaders (the Akkadians) who founded the first Babylonian Empire (Sargon, around 2,300 BCE). In the 2nd millennium BCE, other Semites (Assyrians) dominated Upper Mesopotamia and Syria. In the 1st millennium BCE, other groups of Semites (Arabs) traveled across the Arabian desert. In Africa, the Egyptians in the Nile valley were boxed in between the Libyan and Nubian deserts to the west and the Arabian desert to the east (between the Nile and the Red Sea). The Bible recounts the story of the Hebrew tribes wandering between Syria, Arabia, Sinai and Egypt, the conquest of the "Promised Land" (Palestine) around 1250 BCE and tells of the conflict, even in early Biblical times, between the pastoralists (Abel) and the sedentary peoples (Cain). At the same time, Aryan pastoral farmers from Central Asia, the Medes and then the Persians ruled over the deserts of Iran, Mesopotamia and even Egypt. In the 9th Century BCE, the Phoenicians in the Levant began establishing colonies in Tripolitania and North Africa, and in the 5th Century BCE, Phoenician sailors most probably reached the Saharan coastline along the Atlantic ocean.

In the 4th Century BCE, Alexander of Macedonia (today better known as Alexander the Great) conquered Egypt and founded the city of Alexandria on the coast, along the Nile delta. He went on further into the Libyan desert (322 BCE) as far as the Amon (Sioua) oasis. In the ruins of Carthage (146 BCE), the Roman republic established an African province that would engage in military combat with the North Saharan nomads: the Gaetuli and the Garamantes. In these circumstances, the Roman army carried out multiple raids into enemy territory in the desert, well beyond the *limes* or boundaries marking Roman territory. Through Egypt, which became a Roman province around 30 BCE, the Roman army reached the Red Sea and the First Cataract of the Nile (Elephantine and Aswan). From Tripolitania, the army even advanced (20 BCE) as far as the Ghadamis and Fezzan oases (Phasania). In the 1st Century CE, under the Roman Empire, patrols explored the regions around the Oued Guir (42 CE), once again explored Fezzan and perhaps went as far as Aïr, Niger and Chad (70–80 CE). However, the Sahara of this time was a little less desertic than today and was viewed by the Romans as a strategic space between the Mediterranean and the "land of the Black peoples", offering no direct commercial interest other than allowing the transit of gold or slaves. This region was, therefore, never entirely occupied nor conquered.

Thus, classical Antiquity had a direct experience of the desert through military and commercial expeditions.

2.4.2. *The Middle East*

Along the eastern edge of the Western world, there was little to no knowledge of Asian deserts beyond the Tigris prior to the rise of the Persian Empire (6th Century BCE). It was Alexander's expeditions (331–324 BCE) that first revealed information about these deserts. After having defeated the Persians under Darius III (331 BCE), Alexander entered Mesopotamia (Babylon) and then went on into Persia and occupied Susa and Persepolis (331 BCE). From here, going around to the west and to the north of the Iranian deserts (Lut and the Great Salt Desert), he reached Hyrcania (south of the Caspian Sea), Parthia (Khorasan) and Afghanistan (330 BCE). He then crossed the Hindu Kush range (329 BCE) and entered Bactria, going as far as the Oxus (Amu Darya) and into Sogdia, as far as Jaxartes (Syr Darya) in the deserts of Turkestan. From here, he went to India, descending into the Indus valley along the Thar desert (326 BCE). He returned to the west in 325–324 BCE through Gedrosia (Baluchistan) and Persia. By the time Alexander died in Babylon in 323 BCE, he had conquered all the deserts of the Middle East and central Asia, in 7 years.

Alexander's successors fought over an empire that had extended from Egypt to Sogdia, threatened in the north by the Parthians who had come from Hyrcania in the 3rd Century BCE, and in the west by the Romans. By the 1st Century BCE, the Hellenistic Seleucid dynasty occupied only Syria and Palestine.

Egypt was ruled over by the Ptolemys, descendants of one of Alexander's generals. The Parthians had won all of Iran, Mesopotamia and the upper basins of the Tigris and the Euphrates. The Romans, who were in Asia Minor and Cilicia soon came into conflict with the Seleucid Empire and won Syria, turning it into a Roman province in 64 BCE. However, they were unable to defeat the Parthians, who dealt a resounding defeat in 53 BCE. The Romans could not progress beyond the Parthian Empire into the deserts of Asia and Arabia. From behind their *limes*, running from the Black Sea to the Red Sea, they contented themselves with controlling the starting points (Palmyra, Damascus) of the trade routes toward the East.

2.4.2.1. *The Far East*

At the same time, at the other end of Asia, Chinese civilization was also engaging with deserts. From the 3rd millennium BCE, at the very least, the loess-rich regions to the north and northeast of China, inhabited by sedentary farming villagers, lived with the constant threat of attacks from the Xiongnu "barbarians", who were nomadic pastoralists and skilled horsemen who roamed the steppes of Mongolia and the Gobi desert. These barbarians would harass the villagers through incessant raids and destruction, profiting also from infighting among the Chinese nobility and hegemons. In order to protect themselves, the oldest dynasties (Shang and Zhou) erected isolated fortifications sporadically between the 16th and 3rd Centuries BCE. Toward the end of the 3rd Century BCE, the first founder of the Qin dynasty (220–210 BCE) – the first Chinese emperor to rule over a federation of states that had earlier been independent – joined together these standalone ramparts, creating the Great Wall, running from the Yellow Sea (the Bohai gulf) to the Alashan desert (to the west of the Huanghe loop). Toward the end of the 2nd Century BCE, the Hans, who had become the rulers of China in 206 BCE, extended the Great Wall to the west (between 117 and 100 BCE) and tried, with little success, to carry out expeditions beyond the walls into the southern Gobi Desert. In the second half of the 1st Century BCE, while seeking to form alliances against the Xiongnu, they expanded their sphere of influence, at least periodically, over the Lob Nor depression and the Tarim basin (the Taklamakan desert), as well as over the Turfan oases at the base of the eastern Tian Shan mountains (59 BCE–24 CE). Assured of control over the roads, the Han kings sent out diplomatic and trade missions (and the occasional military mission) to the West, to the Parthians and Bactrians and to Fergana, thus establishing a link to the Roman Empire through the deserts of Central Asia. However, the situation collapsed again toward the end of the 1st Century CE. The Tarim was once again the scene of battles with the barbarians; the routes to the West became more unsafe. And with the Han dynasty falling into disorder (220 CE), the desert was once again won by the nomads.

2.5. Deserts known to Antiquity

The people of the Ancient world were not especially interested in deserts – except, of course, for those who had settled around oases and the nomads. In general, people avoided deserts and even feared them. Travelers from other lands would not risk entering them unless they had some serious reason for doing so and even then, they were more likely to be motivated by profit or power than by curiosity. Further, deserts were almost never considered as such entities in their own right, but were seen instead as a sort of negative of the inhabited world (the *oekoumène*).

Nonetheless, considerable knowledge was gleaned during this period: accounts of travels, trade journeys, military or diplomatic reports and historical or geographic works. They contributed to the development of international relations and the economic growth of different States. For the most part, however, the information stayed confidential or was very poorly disseminated. The information that did reach the public was fragmented, hard to localize and often embroidered with myths or legends. For instance, it is suspected that the accounts of voyages along the African coasts (ships sent out by Pharaoh Necho II in the 6th Century BCE, or voyages carried out by the Carthaginian explorer, Hanno in the 5th Century BCE) contained both accurate information as well as deliberate falsehoods, with the aim of misleading any future competition. Herodotus of Halicarnassus (5th Century BCE), a historian before the concept of a historian existed, recounted the history of the Persians and had visited Egypt himself. In spite of his desire to be objective, his vivid and well-documented account was a mixture of realism and embellishments, containing accurate information as well as unverified facts and extravagant fables. On the other hand, other reports from this period could be dry and fastidious, including reports of various other missions, such as the account of Zhang Kian's Chinese ambassador to Bactria or that of Lucius Cornelius Balbus' Roman expedition to Fezzan (21-20 BCE).

In the West, the most serious documentation on deserts was put together by the scholars who traveled with Alexander (331–326 BCE). Archived in the library of Alexandria in Egypt, this documentation was consulted and used by Greek philosophers and geographers such as Eratosthenes (3rd Century BCE), Hipparchus (2nd Century BCE) and Strabo (58 BCE to 25 CE). The known Western world, as described by Strabo in his *Geography*, lay between India and the Atlantic Ocean and was divided into north and south by the *diaphragm* (lying, approximately, along the 36th parallel). The arid world occupied a large part of this region. In the north, steppes and uninterrupted stretches of desert lay along the shore of the Black Sea, the Hyrcanian Sea (the Caspian Sea) and the Aral Sea and covered Bactria (Turkestan) and Sogdia (Fergana). The Greeks did not know about the deserts further east, beyond the Pamir range. For information on this, one would need to refer to Chinese cartography. In the south, the deserts were broken up more frequently. They were divided into numerous regional compartments, the sphragides: the Indus valley (Thar desert), Baluchistan (Western Pakistan), Persia (the Lut desert and Iranian desert), Mesopotamia, Syria, Arabia and, of course, Libya (the Sahara). The southern border of the known world was the "outside sea" (the Indian Ocean) in Asia, and in Africa it was the "land of cinnamon" (Ethiopia and the Great Lakes Plateau). The Western frontier was the Atlantic Ocean, but what lay inland was a mystery. We only know that Herodotus had heard about the Niger, which he took to be the upper Nile, and that according to him (Herodotus ca. 443 BC–420 BC, Book II, 32) "Beyond the sea and the coast inhabited by men, Libya is filled with wild beasts. There is sand, a terrifying lack of water and a total desert". Toward the end of the 2nd Century CE, the Roman Empire had gone through the deserts from the Atlantic to the Black Sea. In fact, Libya gave the Empire one of its last great Emperors: Septimius Severus, born in Leptis Magna in Tripolitania in either 145 or 146 CE. Starting from northern Africa, caravans would regularly cross the Sahara toward the Niger or Sudan in search of gold, salt, wild animals and slaves. Others would set out from Anatolia or Syria, going through Mesopotamia into Arabia or Persia. South of the Caspian Sea, the Parthian Empire offered a route for caravans from China through Mongolia, Fergana and Turkestan. Nomadic merchants and Greek, Phoenician, Roman, Parthian, Chinese and Mongolian caravans, thus passed from the Syrian desert into the Gobi desert and along the routes through Central Asia, which were the first Silk Roads and Spice Routes.

The Romans, who were more practical than curious, did not undertake any voyages to systematically explore territories and added only a few details to existing knowledge of deserts. Instead, they used the work of Greek scholars to map out the places and routes used by their predecessors. Their geographers were primarily compilers who made use of academic sources, such as Pliny the Elder or Marinus of Tyre (1st Century CE). The most famous of these was Claudius Ptolemy (2nd Century CE) who, in addition to being a mathematician and astronomer, must also be given credit for integrating all available documentation into a depiction of a world system, which was illustrated by a map. His *Geographia*, a major work, was largely a catalogue of places and their coordinates. While it is a work that does contain some errors, it is a compilation of the geographical knowledge of that time. Although, of course deserts figure in the work, there are many uncertainties around them and they are not treated as being of particular significance but simply listed as one of the elements of the Oecoumène. The map that accompanied this text survived and is available to us today only because of the work of Arabic geographers and the copies that were repeatedly made until the Renaissance. Constructed on a canvas and using an original projection, the map covered all of the deserts in Africa and Eurasia from the Atlantic shores to China and spanned 180° of longitude instead of the actual 130° that we know today. This error resulted in Asia being exaggeratedly stretched toward the East, wrongly bringing it closer to Eastern Europe.

While covering less of the world, Chinese cartography from the same era is incredibly detailed but was restricted exclusively to their Empire.

The West gained no new information, in geographic terms, until the end of the 13th Century CE. For over 10 centuries, Western Europe would remain cut off from the deserts of the world following the decline of the Roman Empire, and then through the Barbarian invasions and the infighting among European states. Apart from the Crusades, the only link Europe had with deserts during this time was through large populations of people who came from the arid territories of the Middle and Far East.

2.6. Deserts as corridors of migration

At the beginning of the 3rd Century CE, two political powers, at two ends of Asia, were in direct contact and in direct conflict with the desert and the peoples of the desert.

At the western end of Asia, the Sasanian Persian had replaced the Parthians (224 CE) but had retained their capital, Ctesiphon, on the Tigris. They intervened between the Barbarians and the Roman Empire, but fought over Mesopotamia and Syria with the Roman Empire. To the east, they advanced as far as Tashkent and Peshawar. When at their apogee (5–6th centuries CE), they ruled over Mesopotamia, the deserts of Iran and Afghanistan and part of Turkestan. Then, in turn, their empire was defeated by the Arabs (7th Century CE).

In the Far East, meanwhile, the Chinese Empire rivaled the Roman Empire in expanse and political and economic power. Like the Roman Empire, it had to fend off attacks from barbarians (its neighbors in the Gobi region: Xiongnu, Turks and Mongols) and the Empire was further weakened by internal battles. However, it succeeded in containing the threat of invaders from the steppes and even welcomed them, more or less willingly, into their territory. This resulted in curious cultural exchanges, leading to the creation of a Sino-nomadic aristocracy, whose influence grew after the fall of the Han dynasty (220 CE). This group gave the Sui (581–618 CE) and the Tang (618–907 CE) rulers who established a protectorate over the western Chinese deserts.

The region lying between these two powers, from the Pamir to the Caspian Sea, was a vast space inhabited by nomadic populations prone to conflict, mobilized by events that took place in the arid confines of the Gobi, the Altaï and Lake Balkhash. The causes of these events were complex and are poorly explained: relative overpopulation in a region with poor land and a harsh climate, the search for new pastures, the attraction of arable zones, the wealth of sedentary populations, etc.

The troubles that had long been seen in the Barbarian world were suddenly amplified in the 4th Century CE. Attacks along the borders gave way to large-scale population migration, with people traveling across large distances. This wave of migration began among the nomads, who bred small, remarkably fast horses that occupied the Altaic plateau and the steppes that bordered the Gobi desert (Turks, Mongols). More or less held off by China's Great Wall, as well as by China's policy of establishing friendly relations with them, the barbarians set out toward the West, through the steppes of Central Asia. As they advanced, they triggered a kind of domino effect, pushing ahead of them the hordes and mixed tribes comprising the Huns, Avars and Siberians, who lived nomadic lives between Lake Balkash and the Aral Sea.

By the end of the 4th Century CE, a section of the Huns and Avars from Central Asia had entered Europe through the steppes north of the Caspian and the Black Seas (370 CE), in turn, causing the movement of the Germanic peoples (Goths and Vandals). The Vandals traveled through Gaul and devastated Spain, finally settling in North Africa (429–534 CE), thus cutting off other Western populations from the Sahara. The combined assaults of the Barbarians finally led to the fall of the Roman Empire in the West (476 CE) and the establishment of Germanic kingdoms in Gaul (Francs), Spain (Visigoths) and in Italy (Ostrogoths), none of whom were interested in the arid regions. At the same time, another section of the Huns – the Hepthalites from Turkestan – attacked India, entering the region around the Thar desert (480 CE) as well as the Sasanian Empire in Iran, although they could not destroy it.

These events, which filled the first half of the 1st millennium CE, brought the West no new information about non-deserts and the West also showed little interest in these regions. The only direct contact it had with the arid world was through the Roman Empire in the east or the Byzantine Empire, which battled against the Huns, the Avars, the Vandals and the Persians until the Arabs arrived in the 7th Century. At the same time, ongoing relations between the non-desert and desert regions changed noticeably. Certain motivations died out or weakened, while new reasons for interactions emerged. Given the assaults by the nomads, the States (Byzantine, Persian or Chinese) undertook military operations that were more often defensive than offensive. The decline of the Romans and the movement of the Barbarians led to a decrease in Western trade with the interior of Africa or Asia. On the other hand, a religious wave swept the region, following the emergence of new religions from India and the Near East. Soldiers and merchants, thus, also turned into propagators of faith and religion spread along (and revived) the old trade roads between the three large economic hubs of their time: Byzantium, China and India.

2.7. Deserts: the birthplace of religions

Desert life quite naturally leads to reflection, meditation and sometimes even mysticism. Everything turns thought this way: the beauty and bare horizons, the splendor of the starry skies, the austerity of everyday life and the silence. Many deserts have welcomed religious men and women seeking purity and solitude; many religions have been born in the deserts or been spread here. These religions, which were guidelines for how to live, a means to establish links of solidarity and were sources of faith and hope, collided with social conditioning and political powers. Their rivalries and successes can be retraced, even today, across the Afro-Asiatic desert.

2.7.1. *Polytheistic religions*

2.7.1.1. *Egypt*

Sedentary communities in the Nile Valley in ancient Egypt were devout practitioners of the cult of the dead. The living practiced a polytheism that was inherited from local divinities, which were more or less emblematic of, or subordinate to, the all-powerful Sun God. An official religion of the Sun (Amon) emerged around the start of the 2nd millennium BCE, headed by a powerful clergy. Powerful enough, at least, to defeat, in the middle of 14th Century BCE, the monotheistic reform (a single god, Aton) imposed by the Pharaoh Amenhotep IV (Akhenaton). The Egyptian religion spread, through conquest, into Libya, Nubia, Ethiopia, Sudan, Phoenicia, Palestine and as far as Syria and Mesopotamia. However, in Asia, it came up against the Assyro-Babylonian faiths. From the 12th Century BCE, with the loss of Asia and under repeated attacks, first from Assyrians and then the Persians (from the 7th Century BCE to the 4th Century BCE), it slowly retreated and was confined to a single Nilotic region, disappearing almost completely after Alexander's conquest of Egypt (332 BCE) and the creation of the Roman province (30 BCE).

2.7.1.2. *Mesopotamia, Persia and Central Asia*

Like the Egyptian religion, the religion practiced in Mesopotamia was that of a sedentary community, pragmatic and adhering to the cycles of nature. The farming villagers of the heroic age, Sumerians and Semites of the 4th and 3rd millenniums BCE (Chaldeans), erected temples to sky gods and fertility deities for the fertility of the land. Their influence, however, did not extend very far beyond the middle course of the Tigris and the Euphrates (Babylonia). The Semitic Akkadian invaders (toward the end of the 3rd millennium BCE) and then the Assyrians (the 19th century BCE) introduced the cult of sky gods and agricultural deities into northern Mesopotamia and Babylon, soon assimilated into the Sumerian pantheon, which had a dominant Sun God (Shemsh). The intertwining of political and priestly powers, illustrated by the role of temples, which became both religious and administrative buildings, ushered in a new sort of religion of the State whose glory depended on conquests by the rulers of the

State. The Assyrian religion, however, was unable to make lasting headway in Syria and Palestine – where the Egyptian religion had already spread widely – nor among the Semitic nomads in the Fertile Crescent. Through a series of often violent conflicts, it also collided with the Jewish religion of the Hebrews, until the Assyrian Empire collapsed (7th Century BCE).

The arrival of the Medes (7th Century BCE) and then the Persians (6th Century BCE) marked the entry of Mazdaism into western Asia; until this point, it had been confined to Central Asia and Iran. Mazdaism, revised and reformed by Zoroaster (Zarathustra, 7th Century BCE), is an ancient religion practiced by the Aryan farmers and cattle breeders in Sogdia, Bactria and Khorasan. As a dualistic religion, it was founded upon the basis of the battle between Good and Evil, where Good would eventually triumph. The religion was served by "mages", who were astrologers and soothsayers; it was spread by missionaries in Turkestan and up to Tarim. The religion met with hostility among the nomadic Parthians, but was well accepted in Mesopotamia by the Medes associated with the Neobabylonians (late 7th Century–early 6th Century BCE). The religion of the Achaemenid Persian Empire spread, along with the empire, into Asia Minor, Syria and even the Indus Valley (6th–7th Centuries BCE). Alexander's Greeks and their successors, the Seleucids (4th Century to 1st Century BCE), and then by the Roman Empire did not interfere with the practice of this religion and it survived for centuries, finally becoming the State religion of the Sasanian Persian Empire in the 3rd Century CE, a position it held until it was replaced by Islam (7th Century CE).

Mazdaism had little influence over the nomads of the Aral-Altaic steppes and Siberia, who were mixed populations having in common only their wandering pastoralist and warrior lifestyles. Most of them remained faithful to nature deities – deities of the sky and land – belonging to their traditional belief systems. While they may have been sporadically or transiently influenced by the large religious movements of the time (Zoroastrianism, Buddhism, Nestorian Christianity and then Islam), they were mostly involved in the great military and political conquests of the Seljuk Turks (11th–12th Centuries CE) and then the Mongols (13th–14th Centuries CE).

Unlike the Egyptian and Mazdaist religions, which were closely linked to powerful States but whose spread was limited to the orders of these empires, there were four large religions with a universal calling that reached the arid zones across continents, though at different times and to varying extents.

2.7.2. *Buddhism*

Buddhism is a religion originating among sedentary peoples, whose goal it is to attain, through meditation and asceticism, the liberation of the soul, detached from the material world. It was also a missionary religion that spread through the world, notably along routes of communication, through buildings of worship, and monasteries. Its beliefs were taught for the first time in the environs of Benares, toward the end of the 6th Century BCE and from the 5th to the 2nd Century BCE, Buddhism extended its influence all over India (including Kashmir and the Punjab region) and beyond, as far as Ceylon and China. Traces of Buddhism have been found even in Afghanistan and Turkestan.

Around the start of the Christian era, the original doctrine was divided into two main streams of thought. One of these, the Hinayana (or "Small Vehicle"), spread over the sea into Southeast Asia and Indonesia. The other, less doctrinal and more pragmatic, called the Mahayana (or "Greater Vehicle") spread beyond the Hindu Kush into the deserts of Central Asia. In the West, the spread of Buddhism was restricted by Sasanian Mazdaism, but from the 1st Century CE, it spread eastwards along the Silk Routes toward Tarim, Mongolia and North China. However, it was mainly adopted in cities, oases and centers of trade, along with Mazdaism and Nestorian Christianity. It did not win many adherents among the nomads, except in Tibet where a new form of Buddhism, combining both streams, entered through the Himalayan passes around the 7th Century CE, shortly before Islam spread across the rest of Central Asia.

2.7.3. *Judaism*

Judaism is the religion of Yahweh, the One God of the Hebrews who was revealed to Moses on Mount Sinai around 1,250 BCE. The story of the Hebrews is recounted in the Bible. It is the history of a nomadic people who are gradually sedentarizing, and of the founding of a religion that would spread all over the world. In the first half of the 2nd millennium BCE, the Semitic clans, the Hebrews (Abraham and Isaac), herded cattle between the Chaldees and Egypt across the Syrian and Arabian deserts (Genesis, 12:50). One group settled in Upper Mesopotamia and in Palestine, while another group, called the Israelites (descendants of Jacob, also called Israel), went into Egypt. Here, through the help of one of their clansmen (Joseph), who became the Pharaoh's minister, they settled along the eastern edge of the Nile Delta. Pushed into forced labor the Israelites were delivered from their plight by Moses, who led them across the deserts of Sinai and Jordan, to the gates of the "Promised Land" of Canaan, in Palestine (Exodus). Here, there was accelerated sedentarization in the 12th and 11th Centuries BCE, at the cost of pitched battles with other nomads and the sedentary occupants of the land (Philistines). This period of Judges, war lords and clans preceded the creation of the Kingdom of Israel by Saul in 1030 BCE.

From this time onwards, with the tribe having settled, adherence to the Land and Law of Yahweh led to an intense urban and religious life, marked by the production of the texts of the Bible (11th Century BCE to 2nd Century BCE) and by the construction of the Temple of Jerusalem (around 950 BCE). Toward the end of the 10th Century BCE, following internal rivalries, the Hebrew kingdom split into two: the kingdom of Israel to the north, whose capital was Samaria, and the kingdom of Judea in the desert to the south, whose capital was Jerusalem. There then began a long period filled with strife, combining political and religious conflicts between these countries and also their neighbors, Assyria and Egypt. The Assyrians, who were exasperated with the Jewish prophets preaching against idol worship, invaded the kingdom of Israel (730–720 BCE) and established dominance over Palestine, controlling religious activity. Following an intense Assyrian-Egyptian rivalry (7th Century BCE), experienced in different ways by the Hebrews, Jerusalem was won by Nebuchadnezzar in 587 BCE, marking the end of the kingdom of Judea and the deportation of the Jews to Babylon, until they were delivered from this exile by the Persian king Cyrus, in 538 BCE. This period marked the beginning of the diaspora, the dispersal around the world of exiled Jews and their religion.

The diaspora had grown in strength by the time of Alexander's expeditions. Jewish businesses were set up along land routes over Asia, in Egypt (Alexandria) and throughout the Mediterranean basin. Open, entrepreneurial, as well as missionaries, the Jewish diaspora produced many proselytizers but would often come up against competition or the hostility of other religions. The Jews who stayed back in Palestine, on the other hand, were more withdrawn and, generally, more fundamentalist (e.g. the Essenes of the Dead Sea region) but faced less political pressure, and often changed their relationships with the outside world. While they were not immune to a certain degree of Hellenization during the Seleucid period, they reacted strongly to the dominance of the polytheist Romans after Jerusalem was taken in 63 BCE, the establishment of Judea as a Roman protectorate in 37 BCE and its annexation as a Roman province in 6 CE. Despite the compromises made by the high priest of the Temple (the Sanhedrin), the orthodoxy (the Pharisees) and the conservative aristocrats (the Sadduces), periodic revolts broke out. Rome undertook punitive expeditions against these revolts. The campaign carried out by Titus in 70 CE resulted in the destruction of the Temple and the ruin of Jerusalem. Mortified, Judaism retreated into the synagogues; many Jews emigrated to Transjordania or added to the diaspora in regions as distant as Morocco and Sudan.

2.7.4. *Christianity*

It was in this context of political defiance and moral renovation that there arose among the Jews what seemed, at first, simply one new sect among others but which went on to become a religion with a universal calling. The precursor to this new faith was John the Baptist who, around 28 CE, baptized his followers in the waters of the Jordan and preached in the Judean desert of the coming of "one who was to bring all righteousness". John was the messenger of Jesus of Nazareth, in Galilee, who traveled through cities and villages, teaching crowds in synagogues. He preached the equality of all men, liberty for the oppressed and redemption for all humans, speaking in the language of the people (Aramaic). This new doctrine, a departure from Jewish tradition, scandalized many and, though it won adherents among the poor, was met with hostility among the rich and powerful in Israel, with the result that at Passover in the year 29 CE, Jesus was brought up before the Roman Prosecutor, abandoned by his disciples and crucified.

After the death of Christ, his message of Salvation spread rapidly among the pagans (the "Gentiles") through the preaching of Paul and the Apostles, who were "sent throughout the world to preach the Good News" (the Gospel). The spread of this religion, facilitated by the rise of poverty in the wake of economic decline, had a greater impact in the cities than in the countryside. It was adopted in cities in Palestine, Syria and Anatolia, all the way up to Mesopotamia and India. By the middle of the 1st Century CE, Christianity was present throughout the Roman provinces in Asia and had entered Persia, Egypt and North Africa. This alarmed the Jewish authorities in Judea and the Imperial powers in Rome and set off a wave of persecution that only ended in the early years of the 4th Century CE.

The conversion of the Emperor Constantine (312) and the eventual official recognition of Christianity ushered in a new phase of devotion to the religion of Christ in the East. The end of the 4th and 5th Centuries saw a large-scale religious upheaval across the Byzantine Empire and far beyond its borders. The Church organized itself into dioceses (bishoprics) under the aegis of the "patriarchs" of Constantinople, Alexandria, Jerusalem and Antioch. Bitter theological and doctrinal controversies led to the convocation of ecclesiastical gatherings, synods and councils. Schisms erupted and heresies arose, which had significant political and territorial repercussions.

Arising as a reaction to the Bishop of Carthage, who was considered overindulgent of pagans, *Donatism*[6] did not really spread anywhere but to the African provinces and Numidia, which had been Romanized since ancient times. On the other hand, *Arianism*[7], which originated in Alexandria, spread largely among the Barbarians. Despite the fact that it was condemned by the Council of Nicaea (325), it traveled not only through Europe but also into Asia Minor and Persia.

6 Donatism: a schism that divided the Church from the early 4th Century until the Arab invasion. The name comes from that of Bishop Donatus Magnus, who held that the holiness of the sacraments depended on the purity of the clergy.

7 Arianism: doctrine of the Bishop Arius, priest of Alexandria at the start of the 4th Century. The doctrine specified the relationships within the Holy Trinity, between the Father, the Son and the Holy Spirit.

Monophysitism[8] had proved problematic in Byzantine political life for a long time, arising against a backdrop of disputes within the Imperial power. Traces of this doctrine can be found even today in the Coptic Church in Egypt, Ethiopia, Palestine, Syria and Armenia.

However, it was the doctrine of *Nestorianism*[9], put forth by Nestorius, patriarch of Constantinople, that resonated most strongly with the peoples of the deserts of Asia, at least up to the 12th and 13th Centuries. Condemned by the Council of Ephesus (431) and opposed strongly within the Byzantine Empire, Nestorianism spread eastwards, most notably to the Far East. Although in Egypt, Monophysitism blocked its progress, and in Tibet and India it was blocked by Buddhists, it was well-received by all opponents of the Empire: it supplanted Arianism among the Sasanian Persians, won over Iraq and led to the installation of a Nestorian Patriarch at Ctesiphon. It expanded through Palestine, Syria and Arabia (the Arab kingdom, Al-Hira). Phoenicia even saw the enthroning of a bishop of the nomads (451 CE). Towards the end of the 5th and 6th Centuries, Nestorian missionaries took to the Silk Road, setting up numerous Christian communities along the way, in the cities of Transoxiana, Turkestan and Tarim, as far as Mongolia and China.

The period toward the middle of the 1st millennium CE is essential in understanding how religion functioned and understanding the battles for influence in the deserts of Asia and East Africa during the Middle Ages, and even up to the present. The spread of Judaism, through the diaspora, and of Buddhism through missions, started in the 1st Century CE and continued with varying results. The expansion of Christianity, in all its forms, was especially remarkable, both within the borders of the Byzantine Empire and beyond, around the Red Sea (Abyssinia, Yemen), in Central Asia (among the Huns, Hepthalites and the Turks; the creation of a Bishopric in Merv), and among the monks and hermits in the deserts of Egypt, Libya, Nubia, Arabia and Mesopotamia. There was, nonetheless, a stark contrast between evangelization in cities, based on the basilica church, a structure that was both religious and an object around which sedentary life arose, and the evangelization among the nomad camps, which were more difficult to access and control.

However, all of this would be completely overturned from the 7th Century onwards, following the emergence of Islam. Nonetheless, islands of Christian resistance would continue to exist, despite Buddhism, in North China, until as late as the 19th Century, among the Turks of Amu Darya and the Uighurs of Xinjiang until the 18th Century.

---

8 Monophysitism: a doctrine that refused to accept duality (human and divine) in Christ and stated that he has only a divine nature.

9 Nestorianism: a doctrine proposed by Nestorius, patriarch of Constantinople, which considered Christ to have separate human and divine natures.

2.7.5. *Islam*

Muhammad, the founder of Islam, was a man of the desert, like Abraham and Isaac, and to a much greater degree than Jesus. Born around 570 CE in Mecca, a merchant city in Hejaz, on the Arabian coast of the Red Sea, he spent most of his youth in an Arab family of traveling merchants. A man inspired by visions, but not of divine nature himself, he considered himself to be a "messenger of God" like the prophets, from Moses to Jesus, and he wished to propagate the Word of Allah, the One and Almighty. Muhammad first preached at Mecca, from 614 to 622, where he urged submission to God (Islam) and for people to live a life of virtue and generosity. This simple and moralizing religion, which was viewed with suspicion by rich merchants who were exhorted by the religion to repent and practice charity, was, on the contrary, easily accepted by the nomads and the poor. It thus saw moderate success among the converted: there were those who had faith in God (the Muslims) and there were the unbelievers who were wary of the possible adverse socioeconomic consequences. Persecuted by his detractors, Muhammad left for Medina, where he knew he would be welcomed and find protection. This "emigration" (Hijrah) took place in 622. This is regarded as the first year of the Islamic Era.

Muhammad settled in Medina, a small village in Hejaz, located 350 km to the north of Mecca, and devoted himself to refining his doctrine and organizing the community of believers. He was both a religious prophet and a political arbitrator and intervened in the disputes between Jews or Judaized Arabs (chiefly sedentary peoples), and the nomads, who were more or less pagan or Christian. At the same time, Islam became more radical, the duties of the believers increased and it proceeded to bring about conversions both through persuasion, as well as forced conversion, through military action, with campaigns launched against infidels. With the merchants converting out of interest, Muhammad set out to reconquer Mecca and the *Bedouins*[10] (627–630 CE). By the time of his death in 632 CE, all of the Arab peoples were united within a federal, political framework and in a single faith that was marked by collective prayer, the mosque, fasting in the month of Ramadan and pilgrimage to Mecca.

Muhammad's successors, the "Caliphs", did not have their founder's charisma, nor could they claim to be anointed as Jesus was. Nonetheless, it was during their reign that the various versions of the Prophet's preachings were identified, collected and codified. This made up the Koran (*Al Kitab* – The Book), which was the source of laws and rules for living. However, even within this monotheist and strict religion, which had no clergy and could be interpreted by scholars (Ulemas and Mullahs) as well as legal experts (Muftis), dissent and discord arose around the legitimacy of the guides or those who led prayers (Imams). The orthodox *Sunnis*[11], who were faithful to

10 Bedouin: Arab nomads of the desert.

11 Sunnis: who conform to orthodoxy, *sunnah*.

the first four Caliphs, formed the majority of the community of believers. The *Shiites*[12] believed that only Ali (the Prophet's son-in-law) and his descendants were legitimate leaders of the community. This group of believers settled south of the Caspian Sea, in Iran, Iraq, Syria and Yemen. The *Kharijites*, a puritanical sect, opposed those who maintained the legitimacy of Ali and professed that any believer could be elected as Imam on the condition that he was beyond reproach, surviving to this day in northern Sahara (particularly among the Berbers of Mzab) and along the coasts of the Indian Ocean.

These divisions, along with the added rivalries between the first Caliphs, threatened the unity of Islam and were only overcome by the desire for conquests, which swept through the Arab community following the death of the Prophet. Between the 7th and 9th Centuries CE, Islamization and Arab conquests went hand in hand. From 634 CE onwards, through various wars, Islam spread beyond Arabia, in all directions, with dazzling speed. Omar, the second Caliph, made use of the anti-Byzantine sentiments among the underclasses and the secessionist sentiments among the Monophysits, to conquer Syria and Palestine (Damascus, 635; Jerusalem, 638) and then Egypt (640-642), Libya (643) and Tripolitania (644 CE). Towards the East, he embarked on the conquest of the declining Persian Empire through the occupation of Mesopotamia (Ctesiphon, 640–642) and Iran (632 – Tehran, 642–644). Omar was assassinated in 644 and his successor, Othman, resumed the march toward the East after a period of stagnation and reorganization. This resulted in the decline and fall of the Sasanian Empire (651). Afghanistan was conquered in 654 and Kabul was occupied in 655. Othman was assassinated in turn (656) and was succeeded by Ali, the Prophet's son-in-law, who was also assassinated in 661. The conquests continued under the Umayyad Caliphate, who established a dissident Caliphate in Damascus in 660 CE. Islamization continued into Central Asia: the Arabs reached the Oxus (Amu Darya) in 662 and then passed into Transoxiana (705–715), where they confronted the Turks. In 712, they entered India, going beyond the Indus into the Thar Desert and into the Sindh delta. In Turkestan and Sogdiana, they battled against the Chinese until 751. Islam expanded equally successfully in the West. The first raids toward the Maghreb through Libya, from 647, met with strong resistance from the Berbers. Kharijite dissent did not help matters, with Kharijite communities forming across northeastern Sahara. In 670, the Arabs annexed Tunisia and founded the city of Kairouan. The expansion continued and in 710 all of North Africa had been conquered and Islamized. In 711, the conquerors led the Berbers into Spain and then into Gaul, where they were halted close to Poitiers in 732.

By the end of the 8th Century, all of the deserts in Africa and Asia had been Islamized and were held by the Arabs, with the exception of the Caspian-Aral-Balkhash region, Tibet and Chinese protectorates controlled by the

12 Shiites: those who believed that Ali was the Prophet's legitimate successor.

Tang dynasty (Dzungaria, Tarim, Alashan). In just a few decades, the Arabs had overthrown all former rulers of the Afro-Asiatic deserts. The Western world suddenly found that on their southern and eastern flank, as the Chinese discovered on their western flank, they were now bordered by people whose power and religion had been born in the desert.

2.8. Deserts and empires in the Middle Ages

During the Middle Ages, the arid regions across Africa and Asia were either completely or partially annexed by, or subjugated by, political powers entirely made up of peoples of the desert or steppes. These powers controlled vast areas – indeed, the sheer size of their domains proved to be their vulnerability. Unique local conditions, personal ambitions and religious rivalries precipitated these provinces into dissidence and left their borders open to their neighbors, leaving the carved up empire prey to new arrivals. This was the fate that befell the empires of the Arabs, Turks and Mongols, one after the other, between the 8th and 14th Centuries CE.

2.8.1. *The Arabs*

2.8.1.1. *The Abbasid Caliphate*

In 750 CE, following the battles for influence fought between Muhammad's descendants, the Sunni Abbasid dynasty, descended from one of the Prophet's uncles from Iran, succeeded the Umayyad dynasty. In 762 CE, the newly established city of Baghdad replaced Damascus as the state capital. Arab power extended from the Pamir Mountains to the Atlantic Ocean, however the heart of the empire was still in the arid regions of Arabia and the Fertile Crescent, from Mesopotamia to Syria, and to Palestine.

The first few provinces to liberate themselves were, of course, the most far-flung regions in the kingdom, especially those in the West, which were most prone to rebellion and were the most recently Islamized. In Spain, from 756 onwards, an Umayyad Emir had created an autonomous Emirate in the arid South, whose capital was Córdoba. In the 10th Century, this Emirate rivaled the Caliphate of Baghdad in richness and culture. The situation in the Maghreb was more complex. Toward 800 CE, one of Ali's descendants defeated the Berbers of Morocco and founded the Idrissid dynasty, whose capital was Fès. To contain the agitation led by the Berber Kharijites in central Maghreb, which threatened their authority, the Abassid rulers propped up the orthodox Aghlabid dynasty, from Kairouan, to oppose them. However, at the start of the 10th Century, another Arab dynasty, the Fatimids, a Shiite dynasty this time, who claimed to be descended from the Prophet's daughter Fatima, seized, in turn, all of eastern Maghreb (909).

In 969, the Fatimids conquered Egypt and established a Caliphate in Cairo, which rivaled those of Baghdad and Córdoba. In order to subdue the Berbers, toward the middle of the 11th Century, they deployed a tribe of Arab Bedouins against them, across the desert. The economic and social decline of medieval North Africa is attributed to this tribe, the Banu Hilal.

At the same time (1056) at the other end of the Sahara, another group of camel-herding nomads, Berbers from the borders of the Sahel, spread out across Morocco, where they founded Marrakesh in 1062, then into the west of the Maghreb, and finally conquered Spain (1086). Residing in monastery-fortresses (called *ribat*), the Almoravids (from *al-Murabit*), who were recent converts (converted by merchants and Arab missionaries) to a pure and strict Islam, were fanatical and proselytizing warriors. They would go on to establish a reign of austerity and holy war over all Western Muslims for close to a century, but also ushered in prosperity through trade with black Africa and created the glorious Andalusian civilization. Around 1145, they were replaced by other puritanical Berber Muslims, the Almohads, who came from the Anti-Atlas region, and who would go on to reign over North Africa and Spain until the middle of the 13th Century.

The Asian parts of the Abbasid Caliphate were not immune to these religious and political agitations either, though these were closer to Baghdad. In the 9th Century CE onwards, the dissident Imams protested in Arabia and Yemen. In the 10th Century, a military faction from western Persia extended its "protection" over the Caliphate territory (946). The Fatimids, having captured Egypt, invaded Syria (969). In Iran, to the south of the Caspian Sea, events were influenced by social factors, with the indigenous peoples who had been pillaged and robbed, fighting against the wealthy Arab landowners. Short-lived independent Emirates created by local potentates sprang up in multiple places. The Caliph, who became chief among chiefs ("the Emir of emirs"), only remained sovereign leader in theory and in a spiritual sense, and this declined until the Baghdad Caliphate succumbed to Mongol attacks in 1258.

2.8.1.2. *The Arab order*

While the upheavals caused by the dissent prevented a lasting peace from settling, there was a distinctly Arab religious, economic and cultural order that reigned across the Muslim world in the 9th and 10th Centuries. This period saw a renewal of activity along all the traditional trade routes across the deserts.

In the 12th Century, in Africa, the Saharan holdings of the Almoravids included all of the western Sahara, as far as the Tropics, the Erg Iguidi, the Eglabs, the Saguia el- Hamra and the Hamada du Draa. Their successors, the Almohads, dominated northern Sahara, up to Ouargla, Tunisia, with its salt lakes, and the Libyan coast, as far as Egypt. The Fatimids occupied the Nile Valley, the Nile Delta and the Eastern

Sahara and oversaw the Christian kingdoms of Nubia and Ethiopia. During this time, in the Sahel at the border of the desert and the tropical Savanna, the Black States were created, under the influence of Muslim traders who had visited them from the 8th Century onwards. The Ghana Empire, which had a monopoly on gold, stretched between Senegal and Niger, as far as Tagant and Timbuktu. Between Niger and Chad, several autonomous city-States dotted the Housa region and the Bornu empire extended as far north as Bilma.

Trans-Saharan trade saw the rise of large camel trains, which would travel more or less regularly, led by Muslim traders. At no little risk, these dealers carried Islam and manufactured products from north to south, and from south to north they carried gold, salt, precious woods, furs, ivory and slaves. There was no navigation along rivers, except on the Nile, nor any maritime ports, except those on the Mediterranean (Tunisia, Tripoli and Alexandria) and the Red Sea. The starting point or ending point of any journey was through the "gates to the desert": Marrakesh, Sijilmassa, the Touat oases, Ouargla, Ghadames and Zuila in the north, and Oualata, Timbuktu, Agadès, Kano and Bilma in the south. The Berbers of the Maghreb established relations chiefly with Senegal, Ghana and Niger. The relations that the Fatimids established with Sudan, Chad and the southeastern Red Sea region, as well as the kingdoms of Nubia and Ethiopia, ensured that there was a connection to the upper Congo and the Great Lakes Plateau.

Trade relations in the East, which had been disrupted in Asia by the movement of people and the retreat of the Byzantine Empire, were now renewed, from the 8th Century onwards, between the West (chiefly Byzantium and Venice) and China. Iraq (Mesopotamia) became the central hub of this traffic, where caravans from Anatolia, Syria, Arabia or Egypt, loaded with cloth, metals and fine craft-work, would encounter caravans from China, India or Siberia, bearing silk, spices, precious stones, cotton and furs. Within Africa, Egypt played a role similar to Iraq, for traffic from Mesopotamia, the Maghreb, Ethiopia and Sudan.

Large cities came up along these roads, which went over the deserts or skirted them: Cairo, Aleppo, Damascus, Baghdad, Rayy (Tehran), Ghazni (Kabul), Bukhara, Samarkand and Tashkent. All of these were stopping points as well as major markets. In these cities, you would find artists and traders, either working singly or in guilds or corporations, shops (*fondouks*) and stalls and inns for travelers to rest (*caravanserais*). A vibrant urban and intellectual life flourished along the borders of the deserts and reached its peak in glory, unparalleled to this day, under the Caliph Haroun er Rashid (786–809), the most renowned of the Abbasids and the hero of the *A Thousand and One Nights*. He was famous for his success in Byzantium, and was the ruler of a rich, luxurious and cultured court. The religious and cultural establishments, as well as the libraries, attracted the elites of the Muslim world and drew the attention of the Western world. There were a great number of illustrious scholars and philosophers (some of whose names may have been Westernized) from across these lands. For instance, from Mesopotamia, there was Al Battani (Batenius), an astronomer from Baghdad (858–929), from Iran and Afghanistan there were Razi (Rhazès), a physician and alchemist from Rayy (860–923), and Al Birouni (Aliboron), a traveler, historian, mathematician and astronomer from Ghazni (973–1048), and from Turkestan there was Ibn Sina (Avicenna), a physician and philosopher from Bukhara (980–1037). They wrote in Arabic or Persian, but through the Muslims in Spain, their work traveled to the West and they were renowned all through the Middle Ages. They were guardians of ancient knowledge as well as the precursors to the scientific renewal of the Renaissance.

This bourgeois richness was in stark contrast to the precarious nature of life among the Bedouins of the desert. The sedentary peoples, compelled to look for sources of irrigation, were concentrated around oases, where they cultivated fruits and vegetables, along with cereals, sugarcane and cotton on lands belonging to the bourgeois or to the powerful. The camel-herding nomads, merchants or sheep and goat-herders guarded and gave themselves the right to oversee and dominate these sedentary populations who were scorned and oppressed. This led to an economic and social system that persisted across history till the dawn of the 20th Century in various spots around the arid world.

At the same time, in Europe, Christian States were struggling with invasions from the north and with feudal battles, and had no time to concern themselves with the deserts. They held the Byzantine Empire to be their best defense against the redoubtable expansion of Islam.

2.8.2. *The Turks*

In the second half of the 10th Century, a new wave of unrest and migration made itself felt across the deserts of Asia. The source of this agitation could be traced back to instability among the populations of Turko-Mongols, nomads and warriors from northern Mongolia. For these horsemen, formidable warriors led by elected chieftains, the "Khans", conquering and dominating pastoral lands was necessary for survival. These "Empires of the Steppes", of varying sizes and durability, rose and fell between the 8th and 11th Centuries CE. One of these Empires, the Uighur Empire, was established by conquering the Turkish populations between Lake Baikal and the Altai. The Uighurs, who had a complicated relationships with China, who used them against the Tibetans, rose to glory in the first half of the 9th Century CE. But the Yenisei Kyrgyz vanquished the Empire in 840. Chased out of their lands, the Uighurs sought refuge in China (Gansu), in Dzungaria and in Tarim, where they slowly sedentarized around the oases to the north (Turfan), with the south remaining more or less under Chinese and Tibetan influence.

Thus, around the year 1000, Central Asia seemed to be highly divided and in a state of unrest. China, ruled by the Tang dynasty and then by the Song dynasty, was pushed back toward the East. To the north, Mongolia and the Altai regions remained the domain of Turko-Mongol bands or tribes, who were nomads, polytheists, or Shamanists and were more or less disorganized. The Silk Routes, on either side of Tarim, were controlled in the north by the Uighuirs of Turfan and Gansu, who had adopted Buddhism and Christianity, and in the south by Chinese and Tibetan Buddhists. To the west, the Turks (Oghuz and Karluk) between Balkhash and the Aral Sea, started moving toward the south and further west. Around 990 CE, they entered Iran and Transoxiana (Bukhara was conquered in 999).

2.8.2.1. *The Seljuk Empire*

At the start of the 11th Century CE, a large population of Turks, the Seljuks, originally from the region around the Aral Sea, arrived on the border of Afghanistan. In the years that followed, the Seljuks took Baghdad as their protectorate from the Abbasid Caliphate (1055) and established a Seljuk Empire in Konya (Anatolia), along the borders of the Arab Empire (1070). Coming into contact with Muslim civilizations led to the Turks converting to Islam. However, they retained their warrior nature, which they willingly used in the service of religious orthodoxy and the Princes who symbolized this orthodoxy. Caliphs and Emirs who needed defense or conquests would employ the war chieftains of the Turks (the “Sultans”) as military experts and to organize their armies. In the name of the Abbasid Caliphate, the Turks extended their domination over Iran, Iraq (Baghdad, 1075) and Syria (Damascus, 1075; Antioch, 1084).

Then a complete political and social restructuring in the Muslim world followed: the Turkish Sultans and their warriors granted themselves the right to wage war and collect prebends; Arabic and Persian scholars took over administration; the bourgeois and the merchants divided large land holdings between themselves and became the driving forces of the economy. Despite these transformations, the sense of community arising from religion and the persistence of Arabic-Persian culture masked differences and maintained a certain degree of coherence in the State.

From the 11th Century onwards, however, the State began, in turn, to break up into smaller pieces. The Sultans did not wait long before seizing the reins of power for themselves (using force, when necessary). Although some part of the Abbasid Caliphate survived in Iraq, and the Fatimid Caliphate of Cairo protected itself in the shelter of the Sinai desert, independent Turkish principalities were established in Iran, Syria and Asia Minor. Their capitals, Mosul, Aleppo, Antioch and Damascus equaled, or even surpassed, Baghdad in power. In Aleppo in particular, the Seljuks were able to set up an influential and ambitious Sultanate, that took on the Byzantine Empire in a direct confrontation, and whose aim was to dominate the entire Mediterranean side of the desert, up to Palestine and even Egypt. Seen as more intolerant than the Arabs in their exaltation of Islam and more formidable as warriors, the Turks were a direct threat to the many Judeo-Christian communities along the coast. In the eyes of the Christians in the West, their expansion made it harder to gain access to the Holy Land and to Christ's tomb in Jerusalem. This was cited as the trigger for the Crusades, wars that were waged repeatedly until they finally failed and were abandoned.

2.8.2.2. *The Crusades*

The call for the first Crusade was raised in France by Pope Urbain II, during the Council of Clermont (Clermont-Ferrand) in 1095. The stated goal of the Crusades was to liberate Jerusalem (conquered by the Turks in 1076) and to regain possession of the Holy Sepulchre. Despite the crushing defeat inflicted on the People's Crusade, the Princes' Crusade led by Godfrey of Bouillon, exploited rivalries within the Turks and the relative passiveness of Arab populations. They did in fact succeed in seizing Jerusalem (1099) and then establishing a political power: the Latin Kingdom of Jerusalem (1100) and the Latin States in the East (Edessa, Antioch and Tripoli).

Seven other Crusades, marked by glorious victories as well as humiliating defeats, succeeded the first, unfolding in western Asia, then in Egypt and then in North Africa all the way up to 1270. The cavalry of the West was powerful and was supported by strong fortresses along the borders of the deserts (St. Jean of Acre, Krak des Chevaliers, Beaufort castle, and Kerak castle) but after their initial successes, they soon had to contend with the hostility of the cities and the fury of the Ulemas, the defenders of the faith. The Turks regained Edessa (1144), sparking off a second Crusade, which failed outside Damascus (1148), despite Byzantine support. The crusaders of this period had to reckon with a formidable adversary, the Sultan Saladin of Egypt (1171) and Syria (1174). Saladin's recapture of Jerusalem (1187) and the incessant attacks against the Crusader States triggered the third Crusade (1190), which gave the Crusaders some victories before the inevitable defeat. The fourth Crusade deviated altogether from its objective, seizing Constantinople (1204) in order to create a Latin Empire there, which would last until 1261.

Despite the Crusaders' efforts, the last few expeditions all ended in disaster. Jerusalem, which had been taken back from the Turks in 1229, was forever lost in 1244. With the Seventh Crusade, inspired by Louis IX of France (Saint Louis), the Crusaders fought against the Turks in Egypt (in the Nile Delta) and occupied Damietta (1249). However, they were defeated at Mansoura (1250) and when Saint Louis was taken prisoner, they had to return Damietta in order to liberate the king. Finally rid of the Crusaders, the Turks made use of the fact that the Caliphates had grown weak and extended their empire from Anatolia to Egypt. The loss of Antioch in 1268 provoked a resumption of hostilities by the Christians. An eighth and final Crusade was undertaken, also led by Saint Louis, landing in Tunisia in 1270.

However, the death of the king put an end to the expedition. By the end of the 13th Century, there were no more Latin States along any desert borders in the East.

For the first time since the Roman Empire, the West found itself in direct contact with deserts for over two centuries and yet, neither desired them nor even explored them. They remained confined to the Mediterranean regions (Syria, Palestine, Egypt, Tunisia) and had, thus, gained no new experience or knowledge of the arid world. The fall of the West, however, meant that for some time they were unable to rebuild relations with the Near East, and Turks were in full control of trade routes across the deserts toward Central Asia.

2.8.3. *The Mongols*

In the 13th Century, the deserts of Asia would once again find new rulers. Making use of the vacuum left by the wholesale retreat of the Uighurs, Chinese and Turks, the nomadic Mongols of the upper basin of the Amur river gradually started occupying the region between the Amur river, Lake Baikal and the Altai. At the end of the 12th Century, they began waging war and this was the last and most violent assault by the barbarians against the two great powers of that time: the Arab-Turkish kingdoms to the West and the Chinese to the East. In less than a century the Mongols had built a new Empire that stretched from Korea to Poland and from Iran to China, along the 60th parallel.

2.8.3.1. *Genghis Khan*

From the 11th Century onwards, there were various attempts to unify the nomadic Mongol tribes, but none of these were very successful. The first real unifier of the Mongol was Temüdjin (Genghis Khan), born around 1160. He established his authority over the Mongols, Turks and Tartars in Mongolia and was proclaimed Khan (Chief or Sovereign) in 1206, then launched his hordes of horsemen in all directions. Having secured his Siberian territories, allied with the Uighurs of the Tarim, he crossed the Gobi Desert and took on China. North China, which had been ruled by the independent Jin dynasty, was soon conquered by the Mongols and Genghis Khan took control of the Silk Road passing through the Alashan Desert. In 1211, the Mongols attacked the Great Wall and occupied Peking in 1215. Genghis Khan then turned to the West and entered Transoxiana (1218), attacked the Turks of the Khwarazmian dynasty and seized Bukhara and Samarkand (1219). The Mongols then overturned the Muslim States and Sultanates in the Middle East and Near East and won Afghanistan, Iran (1219–1220), Armenia (1221), the Caucasus region and Southern Russia (1222–1223).

2.8.3.2. *The Mongol Empire*

Genghis Khan's death (1227) did not halt the expansion of the Mongol Empire. His successors, who divided his kingdom between themselves, continued with the conquests. The Jin dynasty were eliminated (1234) and the Mongols then waged war on the Song dynasty in China till 1279. In the West, a legendary winter campaign (1237–1238) across the steppes of Russia and Ukraine gave the Mongols access to the vast North European Plain up to the Oder, and from Hungary to gates of Vienna (1241–1242). The Mongols found progress harder in the deserts of the Near East, which were divided between the Arabs, Latins, Seljuk Turks, Egyptian Mamluks and Turkmen from Asia Minor. Baghdad was taken in 1258 and they threatened Syria. However, the invasion was stopped in Palestine (1260) by the Mamluks.

The reign of Kublai Khan (1260–1294), who was Genghis Khan's grandson, marked the apogee of the Mongol Empire, which now stretched, nominally at least, across all the deserts and steppes of Asia and southern Europe. However, as with the Arab Empire in earlier times, the Mongol Empire was a political body that was far too large, too diverse and too turbulent for a single master to control. The Empire was subdivided into four parts. In theory, its center of gravity was in the East, in Peking, where Kublai Khan stayed. However, the direct authority of the Emperor (the Great Khan) was limited to a large territory that included North China, Tibet, Korea and the deserts and steppes of Mongolia as far as Lake Baikal. A Khanate, called the Golden Horde, covered the steppes of Siberia, from the Aral Sea to the Caspian and from southern Russia, between the Yenisei and the Danube. They would persist in the region until the start of the 16th Century. Another Khanate stretched across the Altai, the region around Lake Balkhash and Transoxiana, as far as Amu Darya. These barbarian territories were ruled over by nomadic horsemen and cattle-herders contrasted with the Khanate in Persia (Afghanistan, Iran, Mesopotamia and Armenia), which kept a highly evolved sedentary and urban Arabic-Persian civilization alive.

2.8.3.3. *Timur Leng and the end of Mongol domination*

The Mongol order was more or less maintained during the 14th Century, until the arrival of a Turkman (Turko-Mongol) from Samarkand - Timur Leng, also known as Timur the Lame. Exploiting the plotting and intrigue between the Mongol Khans, he had himself proclaimed Emir of Transoxiana in 1370. A brutal and tyrannical man, Timur, who was descended from the nomadic Mongols but had been Islamized, chose Samarkand as his capital and claimed to be the restorer of the empire of Genghis Khan. He was a controversial figure, who delivered harsh punishments against the frequent uprisings in protest of his despotism, and his constant military campaigns were more often incursions and pillaging than genuine conquests.

The Mongols, who had been chased out of China by the Ming dynasty (1368), withdrew into Mongolia and their former capital, Karakorum. In 1373–1374, Timur led a campaign in Mongolia with the aim of suppressing the smaller nomadic chieftains, who claimed to be the successors of Genghis Khan. He then turned on Khanates neighboring his domain, the Khanate of Persia (1375–1387) and the Khanate of the Golden Horde, whose Mongol sovereign had advanced into Transoxiana as far as Bukhara (1383–1388), moving with impunity. The steppes around the perimeter of the Caspian Sea (1383) and Russia (1391) were not spared either. In 1393, Timur invaded Mesopotamia and in 1395, he reached as far as the walls of Moscow. After a brief incursion into India (Delhi, 1398), he returned to the West and attacked Armenia and Azerbaijan (1399). However, although he captured Damascus (1399), his progress was stopped in Syria by the Egyptian Mamluks. He then turned to attack the Ottoman Turks in Anatolia (1402) and, in the Far East, the Chinese Ming dynasty, who claimed suzerainty over the Mongols in Transoxiana. Towards the end of 1404, Timur launched a winter campaign against them. However, his death, in January 1405, put an end to this expedition.

Timur's death sounded the death knell for Mongol domination and the last great medieval empire in Central Asia. In 30 years, Timur had devastated all the steppes and deserts of Asia, China, Asia Minor and even the Indus and Caucasus region; he had also curbed the expansionist ambitions of the Ottoman Turks and China. At the same time, he had endowed his capital, Samarkand, with unsurpassed splendor, bringing together (sometimes by force) a large number of scholars and highly talented persons. The glory of this "Timurid Renaissance", which was continued under his successors, could not, however, prevent the inevitable decline of the Empire. Armed groups made up of unemployed soldiers continued to ransack and pillage the countryside and the desert routes. The Ming dynasty shut off China from foreigners, and the precarious nature of any exchange was compounded by insecurity. This precipitated the decline of overland trans-Asiatic trade relations, which would soon face intense competition from the growth of maritime traffic in the Indian Ocean.

2.9. Deserts known at the end of the Middle Ages

At the end of the Middle Ages, a large part of the arid world, the deserts of America and those in the Southern Hemisphere, was still unknown to the people of the ancient world. The deserts of Africa and Asia themselves, had been explored to widely varying degrees. In fact, although the West had shown renewed interest in these regions at the times of the Crusades and threats from the Turks or Mongols, they were still largely indifferent to the arid world, with the notable exception of the Italian merchants of Venice, Genoa or Pisa. The Byzantine Empire itself, showed little interest in the deserts that neighbored its lands, apart from conflicts or very limited trade relations. It was only the Muslims of Spain and North Africa who maintained a faint echo of Arab-Persian culture and its relations with its desert environment.

Consequently, there was very little reliable information and even where it existed, the propagation of this information was not easy. The people of the desert were quite reticent about their environment, either due to ignorance, or as a precaution, or simply due to disinterest. Experience taught them what was necessary and they felt no need to communicate this. As for the conquering hordes, they focused entirely on their campaigns and left little commentary behind. There were, however, small islands of knowledge: the towns and monasteries, with their monuments and inscriptions, their libraries and teachings, their written records and the scholars. However, the fact remains that people still had to want to visit these, and even when willing to do so, had to be able to.

Of course, traveling in the Middle Ages was not an easy undertaking, even though the world has perhaps never seen as much travel as it did in this period. Journeys were always difficult and often dangerous, especially in the deserts, where there were multiple obstacles both natural and human. The desert climate was hard to tolerate: the Sun, the lack of water, the heat and cold, especially in the deserts of Asia where winter temperatures can descend to −30 or even −50°C. Deserts, by definition, lacked everything required for travel: travelers had to carry food, water, and heavy supplies and equipment that weighed down their bags. The journey was usually made on foot or mounted on horses, mules or donkeys and with pack animals: dromedaries or camels, sometimes yak. The loads they carried were limited by the strength of the animals, the alternative being to increase the number of carriers they took along. Further, as a precaution, travelers would only move in groups, following guides. Thus, some caravans could have as many as hundreds or even thousands of animals who had to be fed, grazed and watered. Lodging was hard to come by, spartan where available, spread far apart and quite often a matter of chance: hospitality offered by a nomad camp, caravanserai (often packed) at certain oases, and monasteries, scattered along the paths. Caravans would stop along the way to bivouac or set up tents, or even simply to exchange news and conversation with people they encountered. The journey was made in long stages, traveling from dawn till night, and progress was slow – in fact, some journeys ran into years! In addition to natural dangers, there were the human dangers as well: no safety against raiders, no one who could come to your aid in case of danger, no reliable protection you could depend on, not to mention the linguistic diversity, political vicissitudes and religious antagonism encountered.

Thus, in order to travel in these conditions one had to be strongly motivated and/or have a taste for adventure. How many anonymous travelers must there have been traversing the deserts of Africa and Asia who left no trace of themselves? How many oral accounts or confidences have been lost forever? This is what makes the written records that could be preserved so precious and invaluable.

2.9.1. *Travelers and merchants*

So who *were* the travelers who were hardy and adventurous enough to set out on these desert roads?

War chieftains, regular armies, mercenaries or deserters, and raiders and bandits who were notorious for their exploits. There were also messengers for the Arabs, Persians, Turks, Mongols and Chinese who would travel on horseback, from stage-post to stage-post, along the roads in these empires. There were officials on tours, tax collectors, and officials of the judiciary or administration, authors of confidential reports that were buried within the archives of the State or its rulers.

The Muslim pilgrims and missionaries, seeking sacred sites or propagators of the faith, were less reticent about their experiences. The pilgrimage to Mecca, which, in principle, was mandatory for all, attracted believers from all lands, especially the Arabs in the West. At the same time, Arab or Mongol rulers would send for missionaries to provide religious services, such as the teaching of the Koran, the conversion of infidels or the construction of mosques. Pilgrimages and missionary journeys, carried out since the beginning of Islam, continued and even increased during the Middle Ages, because of the "Mongol Peace" and religious fervor. Some of these pilgrims maintained logbooks (*rihla*) in which they recounted their adventures and described the lands they visited.

Thus, in the early 10th Century (921), Ibn Fadlan of Baghdad, an envoy from the Abbasid Caliphate to the Bulgarians of the Volga, recorded a detailed description in his *rihla* of the land of the Nomadic Turks in the steppes of the Caspian region and in southern Russia, its harsh climate, the interminable summer days, the animals, plants and religions. In the same period, Al Masoudi, also from Baghdad, who was going to Mecca, visited Syria, Palestine, Arabia and the eastern coast of Africa, a center of gold trade. Toward the end of the 12th Century, in the time of the Almohads, Ibn Jubayr, an Andalusian from Valencia, went to Egypt by sea and traveled across Alexandria, Cairo and the lower Nile valley up to the coast of the Red Sea. Having worshipped at Mecca, he went to Medina and traveled across the Arabian Desert, Mesopotamia, Syria and Damascus, before embarking for home from Tyre, on the Phoenician coast. Ibn Jubayr's *rihla* is an interesting resource as, according to many, it served as a sort of guide, a catalogue of places that were useful to pilgrims, along with observations on material and spiritual considerations. The geographical descriptions, however, are quite brief. Even in the 14th Century, the *Mansa* (the lord) Moussa of Mali, who had converted to Islam, traveled across the Sahara and across Egypt to reach Mecca.

Another category of travelers during this time were the ambassadors or envoys of Christian States sent among the "heathen". This was initially done so that the Western rulers could learn about the Mongols, who had a formidable reputation, and about whom nothing was known in the West. Popes and kings delegated emissaries, usually clergymen, who found support among the Nestorians in the various towns and found a warm welcome extended to them by the Mongol Khans. Two among these emissaries have bequeathed us some remarkable accounts.

Giovanni da Pian del Carpine, an Italian Franciscan who was the envoy of Pope Innocent IV, set out from Lyon in April 1243, with the aim of calling for an end to the alleged persecutions of Christians and to negotiate a Christian peace with the Khan. Aged and no longer very agile, he bravely took on a difficult journey that would last over 2 years. After traveling through Poland and Ukraine, he arrived in the land of the "Tartars" (the Mongols) on the northern shores of the Azov Sea. This land was ruled by Batu, the eldest of Genghis Khan's grandsons and the Khan of the Golden Horde. From here, an escort accompanied the monk through the "northern road" (the Caspian region – Aral Sea-lake Balkhash) into Mongolia, as far as the capital, Karakorum. Here, in August 1246, he witnessed the election of a new Khan, Güyük, another of Genghis Khan's grandsons. Upon his return to Lyon in 1247, del Carpine wrote a report of his experiences for the Pope, which was the first serious contribution to the geography and history of the Mongols at a time before they were heavily influenced by the Chinese.

In 1253, another Franciscan, William of Rubruck, a Flemish monk from Saint Louis' entourage, undertook a mission to the Great Khan, by order of the King. This was an attempt at converting the Khan, resulting from misleading rumors about the ruler's sentiments on this subject. When this failed, it turned into a mission to explore the possibility of an alliance, in order to carry out an attack on the Turks from the rear, as they were judged to be more intolerant than the Mongols. Rubruck left Constantinople in May 1253 and from the northern shores of the Azov Sea, followed the same route that del Carpine had taken 8 years before him. Traveling slowly and observing the ways of the "Tartars", it was only in August that he arrived at Batu's court, somewhere on the Volga, where he stayed until September. Having crossed the Don river and then gone across northern Transoxiana, he traveled along the southern shores of Lake Balkhash to the outskirts of Toussaint. The journey then continued into winter, in the extreme cold, across the land of the Uighurs to the north of Tian Shan (Dzungaria) to Karakorum (December 27, 1253). In January 1254, Rubruck was welcomed into the court of the Great Khan Mongka, who had succeeded Güyük in 1248. He stayed at the court until July when, with great reluctance, it would seem, he set off on the return journey, carrying a letter from Mongka to Saint Louis. This time, he took the route along the northern shores of Lake Balkhash and passed through the Mirzacho'l (the "Hungry Steppe"). This journey lasted 2 months, at the height of summer, across the desert, as he describes it, "never resting [...] sometimes [...] in great danger, lacking food and with horses that were exhausted". Rubruck returned to Batu's court along the Volga, where he stayed again through September and October 1254. Finally, he traveled through the Caucasus (the Gates of Alexander), arriving in Georgia, he then went to Anatolia and, finally, after a detour through Cyprus, reached the city of St. Jean d'Acre (Acre) on August 15, 1255. It was here that he wrote his account for Saint Louis. Titled *Travels to the Mongol Empire*, it was a far more detailed and vivid account than Giovanni da Pian del Carpine's report and was an incomparable recording of the land and the peoples of Mongol countries in the middle of the 13th Century.

Among the various travelers through the desert, the best informed were, of course, the merchants, who had fixed stops and correspondents along the trans-Asian trade routes. Most merchants left no written records, but one of them, the Venetian Marco Polo, left us what is undoubtedly the most famous travel journal of the Middle Ages: the "Devisement du monde" (*A Description of the World*), also known by the name "Livre des merveilles" (*The Book of Marvels*), written around 1298. Marco Polo hailed from a trading family, who had established trade relations with Asia going back almost three centuries. In 1269, his father, Niccolò and his uncle, Matteo, returned to Venice after a journey of 9 years to Kublai Khan's court, bearing a message for the Pope. They took to the road once again in 1271, bearing the Pope's response and taking with them the young Marco, aged 17. They departed from St. Jean d'Acre (Acre or Akko) and took 3.5 years to arrive at the court of Kublai Khan in Peking, traveling "through winter and summer, paying no heed to any danger or inconvenience and traversing the deserts over several days and through many difficult regions".

It is not easy to trace the exact route followed by these travelers, as Marco Polo describes both lands he traveled through himself, as well as other lands he knew only by hearsay. At any rate, "what they found along the way" recounted "clearly and in order" includes Iraq and Iraqi cities such as Mosul and Baghdad, Persia and its deserts, Khorasan and Afghanistan, Samarkand (where Christians and Saracens, or Muslims, lived together), the Pamir, Kachgar, Khotan and the Taklamakan Desert, the Lob Nor and the Gobi Desert, Mongolia and the city of Karakorum (close to present-day Ulaanbaatar) "built entirely of wood and earth" and, of course, the city of Cambalu (Peking) and all of China that Marco Polo visited during the 17 years he spent in the service of Kublai.

When returning to Venice by sea in 1295, Marco Polo was captured by the Genoans and imprisoned until 1299. While he was in prison, he dictated the memories of his travels to a fellow prisoner, Rusticello de Pise, resulting in the *Devisement du Monde*. This account, which was published for the use and edification of the nobility and served as an advertisement of sorts for traders, is remarkable above all for the numerous facts its contains, the accuracy of the observations, the realism of the descriptions and the author's critical thinking, despite the digressions into legends that were common to all books of this time.

There is yet another category of travelers we must name: those who could be called "tourists" along the lines of Herodotus. These were travelers, like Ibn Battuta, who "enjoyed reflecting on curiosities in the cities and the marvels seen on travels", and who enjoyed meeting people and engaging in conversation with them. Ibn Battuta is undoubtedly the archetype of this kind of traveler or, at any rate, the most illustrious. Born in Tangiers, in Morocco, in the time of the Marinid dynasty, he spent a large part of his life on the road chiefly motivated by wanderlust, to satisfy "his heart, which was set on traveling". However, there were also other reasons: The Hajj or pilgrimage to Mecca; observing how the Prophet's laws were applied; seeking to contact scholars, officials and those in power; and exploring unknown places and observing the mores of their inhabitants.

Setting off from Tangiers in June 1325 on pilgrimage to Mecca, Ibn Battuta would only return to Fez in 1349. In his years away he traveled North Africa, Upper Egypt, the Near East, Iraq, Persia, Arabia, the eastern coast of Africa, Asia Minor, southern Russia, Central Asia, India, Ceylon, Sumatra, South China and made four other journeys to Mecca. His return to Fez was brief and he set out again in 1352 to visit Andalusia and then traversed the Sahara via the Erg Iguidi and Mauritania, visiting Mali and Niger. Recalled to Morocco by the Merinid king (his second crossing of the Sahara, this time via Hoggar and Touat), he dictated, to Ibn Juzayy, a letter from Grenada. This was a long *rihla* and is a unique document on the geography of the Arab world in the 14th Century, especially on the deserts of Asia and Africa: Arabia, Persia, Turkestan, Sahara. A complex person, someone both austere and epicurean, censorious but a seasoned courtier and curious as well as verbose, Ibn Battuta (or Juzayy) is a remarkable narrator, who is able to perfectly highlight notable episodes during his journeys, whether anecdotal, funny or dramatic. His documentation is rich and generally seems truthful, although, like Rubruck and Polo, his contemporaries suspected him of sometimes lying. However, the account also remains incomplete, given that Ibn Battuta, who was deeply pious and spoke only Arabic, spoke only to fellow Muslims and lawyers.

We can also add Al Hassan ibn Muhammad al Fasi to this category of travelers. Better known by his title Leo Africanus, he traveled across Africa in the first half of the 16th Century. Born in Grenada in 1483, he was a Moorish refugee in Fez. He traveled to Sahara and Sudan in 1513–1515 and during this time visited Mali, Bornou and Chad. From 1517 to 1520, he traversed Egypt from Cairo to Aswan. As he was returning to the Maghreb, he was captured, close to Djerba, by pirates and then settled in Rome under the protection of the Pope and converted to Christianity, adopting the name Leo. It was in Italy that he wrote and published an Arabic account titled, *Description of Africa* (1526), which was translated into Italian (1550) and then into French (1556) and was consequently widely disseminated in the West.

2.9.2. *Medieval geography of deserts*

2.9.2.1. *Asia and the Silk Road*

In the 15th and 16th Centuries, as the Arab empires and then the Mongol empires started declining and breaking up, Asia saw the rise of independent political entities. The deserts would never again be united under a single ruler.

The Arabian Desert in western Asia, a true desert, which was divided by the Red Sea and surrounded by less arid regions, was ringed by vast oases along large rivers: the Nile (Egypt) to the west, the Tigris and the Euphrates (Mesopotamia) to the east. The most significant event here was the extension of the Ottoman Turk Empire, which would endure until the 20th Century. Toward the end of the 14th Century, this empire extended over Turkmen principalities in Asia Minor and in parts of Europe south of the Danube. Interrupted by the rise of Timur, Ottoman progress continued after his death (1405). Although the expansion was slow at first, it accelerated after the conquest of Constantinople (1453) toward Crimea and the Russian steppes. At the start of the 16th Century, the Ottoman Turks continued to expand, driving back the Mamluks of Egypt in Syria (Aleppo, Damascus, 1516) and going up the Nile (1517). Sparing the blazing desert in the Arabian peninsula, the Ottomans occupied western Armenia, Mesopotamia (Baghdad, 1534), the southern shores of the Persian gulf and set up a protectorate on the eastern coast of the Red Sea, which had the holy cities of Islam: Mecca and Medina. In the East, however, the expansionist ambitions of the Ottomans collided with both military and religious resistances from the Safavids, a new dynasty which, in 1502, eliminated the last of the Mongol princes in Iran. A constant war raged between this Shiite dynasty and the Sunni Turks over Mesopotamia and the Caucasus region. Rulers of a rugged mountainous region with arid plains (the Salt Desert, Lut) "where there is no river, no stream, no water or very little, and what there is salty [...] neither are there wild beasts, who find nothing to eat as the land is hard and dry and no grass can grow" (Marco Polo), the Safavids, who had a well-deserved reputation for art and intellect, dominated Persia from Afghanistan until the Caspian region until the early 17th Century.

Central Asia, from the Caspian Sea to Lake Balkhash, form a corridor among the steppes, interrupted by true deserts (Karakum, Kyzylkum), which are freezing in the winter and blazing in the summer. The Timurids, successors of Timur, were still ruling in these regions in the early 16th Century, in Khorasan and Transoxiana, in the few remaining Khanates and cultural sites of great renown such as Bukhara, Tashkent and Samarkand, "[a] very noble and grand city" (Marco Polo). Driven back by the Uzbeks of Turkestan and threatened by the Safavids of Iran, the Timurids retreated to the South, to Afghanistan (Kabul, 1504), the Pamir Mountains and India (1522), where one of them founded an empire that came to be known as the great Mughal Empire, which included the Thar Desert and the Baluchistan Deserts and which endured until the 19th Century.

In East Asia or Upper Asia, there are vast deserts. There is the Taklamakan, "a region that was almost all sand", the Gobi, which is "so long, that it is said that if one travelled for a year, one would not reach its end and so wide, that at its narrowest, it would take a march of at least one month to cross it" (Marco Polo). There are also wide steppes between massifs that are hard to negotiate and where the cold is extreme in the winter months. This was the historical heart of the Mongol lands. The neighboring lands of North China, devastated by the Mongols in the 13th Century, were taken over by the Ming dynasty in 1368. This dynasty set about re-establishing Chinese peasantry and created large-scale irrigation projects in the deserts of "interior" Mongolia, beyond the Great Wall. To the north of the Gobi Desert, "outer" Mongolia, which had been neutralized, once again became aggressive in the 16th Century, profiting from troubles that irrupted within Chinese provinces. In 1550, the Mongols arrived as far as the gates of Peking. A treaty signed in 1571 established a precarious peace, in exchange for their vassalage, which lasted until the fall of the Ming dynasty (1644) and the advent of the Manchurian Qing dynasty, who ruled until 1912.

The major axis, the dorsal spine of this longitudinal succession of deserts in Asia was the network of overland routes, used for millennia, which have been called the "Silk Roads" since the late 19th Century (Volume 3, Chapter 3, Figure 3.20). This road, in use since Antiquity, and even Prehistory, made the best possible use of difficult terrain: going across the steppes, the piedmont glacis, through narrow mountain passes and skirting the deserts. In the Middle Ages, silk, spices (pepper, cinnamon), remedies (camphor and rhubarb) and porcelain from China, furs from Siberia and Mongolia, wood, precious stones and cotton from India, incense from Arabia – would be exchanged in Syria or Constantinople for metals, amber and linen cloth or wool from Europe. There were two principal routes, one from Crimea and the other along the Mediterranean coast, which intersected at Transoxiana where, in Bukhara, Samarkand and Tashkent, merchants from Venice, Arabia, China and India would all mingle together. From here, through Fergana and the Pamir Mountains, the road went to Kashgar, a port in the Tarim basin, from where "many merchants set sail, going out to do business around the world" (Marco Polo). The road then again divided: one route went to the north, passing by the Tian Shan piedmont range and the Turfan oasis, another going south via Khotan and the southern edge of the Taklamakan Desert. It finally converged one last time at Ansi, to the east of Lob Nor, before breaking up into multiple routes leading to Mongolia and China.

The history of this road is made up of periods of intense activity, punctuated by troubled periods with a total absence of traffic for varying lengths of time. Among these periods of activity along the road, there are the periods of Hellenistic influence, from the 4th Century BCE to the 1st Century BCE, the time of the Arab Empire under the Abassids, from the 8th Century CE to the 9th Century CE and, above all, the time of the Mongol Peace when Venetian commerce ruled supreme, in the 13th and 14th Centuries. On the other hand, the route was cut up and made hard to navigate, during the large-scale barbarian migrations of the 4th and 5th Centuries CE, during the Crusades, and during Timur's invasions in the late 14th Century. There was a sharper decline in the use of these roads in the 15th and 16th Centuries, as it was controlled by the Ottoman Turks in the West, and by the Ming dynasty in the East, who had closed off China, preventing foreigners from entering. At the same time, the Portuguese arrived in India using maritime routes and the manufacture of silk developed in Europe, both of which reduced Western interest in this difficult and precarious overland route. While the road remained open and frequented, it was now largely used by restricted local or regional traffic.

2.9.2.2. *Africa and the roads for the salt, gold and slave trade*

Although it had been in direct contact with Europe since ancient times (Alexander's expedition, Roman incursions, Byzantine establishments and Genoan business along the coast), the Christian West had limited knowledge of Saharan African until the end of the Middle Ages. Less threatening and less tempting than the Asian deserts, the Sahara (with the exception of Egypt) had never really attracted the attention of the Europeans, even when it was involved in events that took place along its Mediterranean borders. For a long time, its history was only passed on in oral traditions and even after the *rihla* of Arab travelers, the paucity of texts on this subject is in stark contrast to the rich documentation of the Asian deserts, and there remained significant gaps in knowledge. The Europeans were interested in the products they could obtain from sub-Saharan Africa (gold and, above all, slaves), not in the desert itself. Here again, the greatest source of information were the merchants, especially the Arabs. The Arabs had traded along the coasts of the Red Sea and the Indian Ocean since at least the 1st or 2nd Century CE and had undoubtedly been taken further inland by African traders. However, from the 7th Century onwards, they established a monopoly on trans-Saharan commerce, trading for themselves and on behalf of the Genoans or Pisans, who rarely risked traveling across the desert. Along with their wares, the Arabs also carried Islam, which spread rapidly, leaving behind a few pockets of Judaism along the oases and a surviving Christian tradition in Nubia and Ethiopia.

To the Europeans, the Sahara had always been "uninhabitable" land, which was nonetheless inhabited by sedentary cereal farmers in the oases, and by nomadic pastoralists who herded goats and camels. There was frequent conflict between these groups and the Black farmers (Haratin), who were pressured by the Berber nomads, often employed the propagators of Islam, such as the Almoravids, against the nomads. The oases occasionally also supplied dates and, when needed, the nomads would hunt antelope. Politically, the Saharans were organized into independent confederations of tribes or villages, or depended on rulers or States located along the borders of the deserts.

On the whole, North Africa, at least through Islam, remained a dependency of the East. In Morocco, the Berber Merinids succeeded the Almohads in 1269. They settled in Fez and Marrakesh and controlled the ports along the western edge of the Sahara, including the most illustrious of these, Sijilmassa, founded in Tafilalt in 758 CE. Other Berbers ruled in Tlemcen, threatened by both the Merinids and the Spanish. In 1229, in Ifriqiya (Tunisia), a viceroy of the Almohads founded the Hafsid dynasty, which would endure until 1574. They had a negligible influence on the Sahara, despite the fact that they controlled Ouargla and the Chotts region. Libya is vast, however its limits are not well defined and its territory is divided. To the west, Tripolitania, which is entirely Saharan and was under the influence of Tunisia, became independent in the 14th Century but was eventually conquered by the Turks after the occupation of Tripoli (1551). To the east was Mediterranean Cyrenaica, Islamized in the 11th Century and disputed territory between Egypt and the Hafsids.

To the extreme east, the Shiite Fatimids who had ruled over Egypt and Syria and had constantly been at war with the Sunnis in Baghdad, the Byzantines, the Crusaders and the Seljuk Turks, were finally overthrown by Saladin in 1171. However, in 1250, the Mamluks, slaves and mercenaries in the service of the Turks, gave themselves an independent Sultan. The new State defended Egypt against the Mongols, but fell, in turn, to the Ottoman Turks (1517). It was the Turks, finally, who between 1529 and 1574, and despite Spanish intervention, occupied Algiers, Kairouan, Tripoli and Tunisia, one after the other, and thus gained control of the entire Mediterranean coast along North Africa, holding this until the 19th Century.

On the other edge of the Sahara, which was called the Bled es Sudan (or Land of the Blacks) there were again confederations or "Empires" that spilt over into the arid lands. The oldest of these, of which we can find mention, is Ghana, which was spoken about in the 8th Century by a certain Al Fazari. This Ghana (not to be confused with the present-day country of Ghana) included the Senegal river basin, the upper Niger and Mauritanian Sahara as far as Tagant. Defeated and converted to Islam by the Almoravids (1076), the country was conquered again in the 13th Century by neighboring Mali. Mali, which was also converted by the Almoravids (after 1050), succeeded Ghana in prestige and power because of their gold mines and trans-Saharan trade. The Empire reached the apogee of its glory in the early 14th Century: its territory at this time extended from the Guinean forest to the desert and from Lower Senegal to the Middle Niger. Mali's munificence was legendary: one of its kings, *mansa* Moussa, set off on pilgrimage to Mecca (1324) and took with him 500 slaves and 100 camels loaded with gold; he lavished these riches upon Egypt and brought back a group of scholars who would cement the intellectual reputation of Timbuktu. Ibn Battuta, who passed through Mali shortly after this, was stunned by the opulence of the Empire and stayed for 2 months to praise the "Sultan" and his people. The prosperous kingdom of Mali was, naturally, coveted by its neighbors such as Tuaregs Mossis and Songhai. It began declining in the 15th Century and its last kings died out in the early 16th Century.

To the east of Mali, a Sudanese Empire had been founded in the 7th Century on Songhai land, on the middle Niger, and its capital was Gao. Sometimes an independent State and sometimes a vassal State, by the 15th Century the Songhai managed to impose their rule on the ruins of Mali and even seized Timbuktu (1468) and Agadez (1515) from the Tuaregs. However, internal fights and trade wars with the Saadi dynasty of Morocco led to the fall of the empire (16th Century). Even further east, the Hausa had formed a federation of isolated States dating back to the 10th Century and that extended from Niger, upstream of Gao, up to Kano and Agadez. This confederation endured until the 19th Century. The Kanem-Bornou Empire, mentioned by Al Bakri in the 11th Century and Islamized in the 12th Century, spread out around Lake Chad, in central Sudan. This was the only empire that was comparable in power to the large States in the West and it also endured until the 19th Century. Finally, beyond this to the East, along the upper Nile, in Nubia, a complex mix of Egyptian and Christian traditions forged over millennia, finally succumbed to Islam (16th–17th Centuries).

The Sahara was barely populated. On the other hand, numerous paths ran through it, upon which people were constantly moving. Unlike the routes of Asia, which followed a long longitudinal corridor, the routes through the Sahara were arranged in internal circuits and meridian routes. Apart from the random movements of the nomads, the internal circuits also saw the movement of other products that traveled from one oasis to another (salt, dates, cereals), required as supplies. The trans-Saharan routes are ancestral routes that connected African producers with Mediterranean consumers and were traveled by extensive caravans, most often led by Berber Muslim guides. Unlike Asia, these routes would not pass through large cities that served as stopping points. They connected the "gates of the desert", located at the ends of the road and which were hubs of transactions, and of business and warehouses run by the Syrian, Jewish or Arab merchants. To cross from one end of the desert to the other, one had to cover, at some peril, enormous distances without water and often without grazing lands. The journeys were interminable. Ibn Battuta, who set out from Sijilmassa, took 25 days to arrive at Teghaza, going across the Erg Iguidi. It then took him another 20 days, traveling 700 km, to travel from Taudenni to Timbuktu.

The routes that are followed have been the same since Antiquity. To the west, there are two roads that connect Morocco and Ghana-Mali-Songhai. One of these starts from Marrakesh or Guelmin, traverses the Seguiet el Hamra (Rio de Oro) and the Zemmour, passes through the Mauritanian salt mines of Idjil and Adrar and reaches either Atar and Senegal, or Ouadan, Oualata and the Niger. The other road starts from Sijilmassa, whose splendor and gardens were praised by Al Bakri in the 11th Century, but which is, today, a mass of misshapen ruins. This road joins Taudenni through the Erg Iguidi, the Eglabs, Teghaza and the Erg Chech, or through Touat and the border of the Tanezrouft, leading either to Oualata in the Hodh or Timbuktu on the Niger. At the center, from Tunisia, starting from Ouargla or the Chotts, one can reach Ghadames, Ghat, Dhanet, Iferouane, Agagez and Gao, and from Libya, starting from Zuila, one can reach Fezzan, Mourzouk, Bilma and Chad or Kano in the Hausa lands. Finally, to the east, the Nile has connected Egypt, Nubia, Ethiopia and the Great Lakes Plateau for millennia.

The chief trade here is of salt, a staple for survival and the only export product made in the Sahara itself. Large salt mines are found in the west: Idjil, in Mauritania, mentioned from the 6th Century onwards, Teghaza and Taudenni along the Niger routes. Other mines are located toward the center: Tekadda (Teguidda n'Tessoum) on the road from Agadez to Gao, and Bilma on the eastern end of the Erg in Ténéré. Salt, which was bartered for cereal and cloth in the Maghreb, was exchanged for gold in the Sahel. Twice a year, in spring and autumn, Mores and Tuaregs would allocate a caravan (the *azalaï*) for transporting salt. Consisting of several thousand camels, this caravan is a tradition that continued until the middle of the 20th Century (see Volume 3, Chapter 3). Caravans also transported gold from Guinean mines in Ghana and Mali, which, until the 16th Century, were almost the sole suppliers of gold to Mediterranean powers. At this time, Timbuktu and Gao had a monopoly on the market. This declined with the use of the Atlantic Sea route by the Portuguese from 1450 onwards. The death blow to the overland trade of African gold was dealt by the discovery of American gold and its spread to Europe towards the late 16th Century. The slave trade continued longer – though it officially ended with the prohibition of 1815, in reality it continued until the middle of the 20th Century, and a clandestine trade undoubtedly continued even longer. Other merchandise, in smaller quantities, complemented this caravan trade. These were, from the south to the north, ivory, the skin of wild animals, ostrich feathers, millets and so on; from the north to the south, there was barley, wheat, oil, cloth, metals and manufactured goods.

In the 14th Century, prosperity from trade flourished in the Sahara and there was large-scale caravan traffic. This even attracted merchants from Venice and Genoa, whose trading in Asia had been disrupted by the expansion of the Ottoman Empire. In the 16th Century, the ruin of the gold trade, as well as political decline in Mali and the Songhai Empire, considerably slowed down mercantile activity in the western Sahara. It continued for a little longer in the east, between Libya, the Hausa lands and Kanem, as well as along the Nile and the Red Sea. However, like in Asia, it would soon depend on the control established by the Ottoman Turks along the Mediterranean coasts.

2.9.2.3. *The representation of deserts*

Medieval geography did not provide any special images of deserts. However, it included deserts in its general representation of the Earth, with noticeable differences depending on the place and time.

All through the Middle Ages, the only knowledge Europeans had of the desert was what certain erudite monks were able to glean from Ancient texts, especially Ptolemy's work. Unfortunately, they allowed themselves to be led into theological speculations that were incompatible with objectivity. For example, there are the "images of the world" (*Orbis terrarum*) called "T-and-O" maps, where the Earth is represented as a circle, surrounded by the ocean and symbolically divided into three (like the Holy Trinity), by two arms of the sea forming a T, with Jerusalem in the center, Europe to the left, Africa to the right and Asia on top, with the East being situated here, the seat of earthly Paradise.

Where commentary exists, it rarely takes into account newly acquired knowledge. The arid regions of Asia were more poorly known than in Antiquity. The large deserts were located beyond Eden, populated by monsters, beasts and serpents. Even in the 18th Century, despite all the travels among the Mongols and African contacts, the deserts were relegated to the rank of "*terrae incognitae*" (unknown lands) or, at the least, uninhabitable lands. On the other hand, legends and mysteries were given great importance and were often accredited by travelers themselves. For example, there was the legend of the Christian Kingdom of Prester John, which was sometimes said to be in Ethiopia, at other times in Iran and other times in Mongolia, and which was likely to have only been a distorted reflection of the Coptic church or Nestorian influence in Central Asia and in the Far East. Or there was the pagan land of Gog and Magog, hidden away behind China's Great Wall or, again, the legend of the Dry Tree or the Solitary Tree, which marked the entry to heaven and to unknown lands, and which was supposed to be located on the eastern borders of the Turkestan deserts. These images, embellished and illuminated by faith rather than reason, would endure until the dawn of the Renaissance in Europe.

In this same period, from the 12th to the 14th Centuries, in Persia, Transoxiana and India, there were remarkable developments in astronomy and mathematics; China saw a host of new inventions including paper, woodblock printing and the compass. Commerce, administration, large-scale projects and missions could give Eastern geographers a wealth of new observations. Chinese cartography, which had largely centered on the Empire, progressively expanded, from the 3rd or 4th Century onwards, to represent all of Asia. This included the deserts, of course, but the cartographers seemed to not know the Western world or Africa, who, in turn, seemed equally ignorant of China.

Nevertheless, Western populations did have access to Marco Polo's books and some even had access to Arab sources. The Arabs were able to maintain continuity between Antiquity and the present because of their translations of Greek authors and could serve as a link between the East and the West due to their overland and maritime relations with Central Asia, India and China. Not only did they pass on Ptolemy's work (the *Almagest*) in its entirety, but they enriched this with knowledge from India and China, as well as their own discoveries. Andalusian and Maghreb geographers played a fundamental role in this field and it is to them that we owe the most complete picture we have of medieval deserts. Thus, in the second half of the 11th Century, Al Bakri, born in Córdoba and who had traveled in North Africa, Sudan and Arabia left behind a *Geographic description of the world* (1068) of which only fragments have survived to our times. In the 12th Century, there was Al Idrisi, who was born in Ceuta around 1100 and studied in Córdoba, with a brief career in Africa, in the service of Roger II, the Norman king of Sicily. In 1154, he constructed, for the King, a large *Mappa mundi* in silver, accompanied by 70 detailed maps from Scandinavia to the Sahara and from western Europe, to Central Asia, to India and China. As was the case with all maps of this time, the Earth (the three continents) is surrounded by an ocean, but as with Chinese maps, in his maps, the North is at the bottom of the map. Geographic representation owed a great deal to Ptolemy, especially the extension of Asia to the East, which so fortuitously misled the future discoverers of the Americas. We can see in these maps the confusion between the Niger and Nile, which dates back to Herodotus, and which would be repeated by Ibn Battuta who had even canoed on the river. But Idrisi added a few new details, especially to North Africa and Asia Minor, which he had visited himself. The commentaries on the map, titled "Diversion for whoever wishes to travel the world", could be considered the sum of Arab geographic knowledge in the middle of the 12th Century and this was widely distributed across the West.

These details would later be included in European cartography. In the 14th Century, maps were drawn up by adding, with varying degrees of accuracy, knowledge from contemporary texts to knowledge inherited from ancient science. In the 15th Century, the large school of cartography, established in Majorca by Arab and Jewish geographers who had been driven out of Spain, included in its work not only the information provided by Marco Polo and Ibn Battuta, but also information collected from Israelite traders from the oases. In 1492, the famous globe created by Martin Behaim in Nuremberg, defined the line of the Mediterranean coasts and the near Atlantic, taking into account information from the *portolans* and observations by navigators. In 1526, Leo Africanus, in his *Description of Africa*, spoke of Djenné, described Timbuktu in glowing words, claimed to have seen the Niger flow west, and described Bornou, Chad and Egypt. He spoke of the existence of a "vast sea" (lake), which could have been the marshy regions within the Niger delta, unless he was speaking of Lake Chad. In the 16th Century, the world maps or *Mappa mundi* created by Mercator (1569) and Ortelius (1570) corrected Ptolemy's longitudinal error and sketched out the earliest contours of the American continent. From this time onwards, the tasks at hand were chiefly those of verifying locations, eliminating errors and adding new discoveries.

Even with all these documents, contemporary people could only extract fragmentary knowledge (not always reliable) about deserts. Interest centered chiefly around humans, their religion, their mores, their societies, their riches and their cities, rather than on their environment. The marking of locations was uncertain, and the vague toponymy made it hard to identify regions, especially when comparing these maps to present-day maps. For instance, the name "Caucasus" is often applied collectively to the mountain ranges in Central Asia that go from the Black Sea to the Himalayas and to the "Eastern Ocean" (the Pacific). Sometimes an author would describe something by hearsay (Tibet, the Gobi Desert, China, many places in the Sahara), and sometimes the authors would dwell at length on personal experiences. Rubruck dedicated a long section to Karakorum, where he stayed for a long time. It was a city or camp that "was not even as large as the city of Saint Denis", but was packed with Chinese and Arab merchants, even Christian artisans, and where he encountered a woman from Metz and a Parisian goldsmith. There were large gaps and few physical descriptions. The most pertinent of these spoke of the weather, the cold (the frozen herbariums on the Volga delta in winter; summer snows on high peaks), the hail and the wind. Descriptions also focused heavily on rivers, their immense sizes (e.g. the Volga, "four times as large as the Seine") and the role they played as frontiers (the Tanais or the Don, "separating Asia from Europe, as the Nile separates Africa from Asia". However, there was very little on the flora and fauna, and almost nothing on topography. At this time in history, we were still far from having any comprehensive knowledge of deserts.

2.10. To be continued ...

I will pause in this overview of the discovery of deserts from the earliest days of the human race, because from the 16th Century onwards, the world would develop a new form of interest in this arid world.

Up until now, deserts had been considered as sort of neutral spaces that one simply passed through, as one would travel on the seas, without stopping anywhere. The disparate social groups and the absence of a unifying patriotic sentiment prevented the formation of internal political units, except along the large rivers in Mesopotamia, Egypt and Mali. The people who lived here (nomads and people of the oases) were most often left to themselves or were vassals of powerful neighbors such as the Arabs, Mongols, Turks, Incas or Aztecs. From the 16th Century onwards, however, deserts would become sought-after, disputed, divided, annexed, colonized and exploited by powers that were at varying distances from the deserts, but were all politically or economically strong and increasingly dominant. Indeed, deserts rarely attracted people for what they were themselves: beautiful and solitary regions. The austerity of these spaces may have stimulated religious ardor and certain forms of modern tourism also highlight these unique qualities. But on the whole, deserts are looked at in a more prosaic and uninterested manner.

Across the centuries, deserts have remained arenas of adventure: from the armies of Antiquity to the conquerors and merchants of the Middle Ages, to the colonists of South Africa and Australia and the Pioneers in Western America. In the 18th Century, the spirit of adventure was complemented by a thirst for discovery and scientific research. The 19th and 20th Centuries saw grand overland expeditions, either by individuals or groups, across all the deserts of the world, either initiated by or supported by enlightened individuals (Napoleon in Egypt), scholarly societies (the Paris Academy of Sciences and the Society of Geography, the Royal Society of London, the National Geographic Society of Washington), or large engineering or technological projects (the American railways, the Trans-Siberian railways, the United States Geological Survey, the cartographic branches of the French and British armies) or financial bodies (mining consortia, oil companies) and so on. Certain sites were used, and are still used, for specific purposes as a result of the advantages they offer, or their availability: for how isolated they are (military nuclear experiments in the Sahara, Nevada, Kazakhstan, Rajasthan and western China), for the purity and transparency of the air (astronomical observatories such as Mount Palomar in California, Cerro Tololo in Chile), or simply because of their natural beauty (filming locations: Monument Valley in Arizona, the South Moroccan Desert). Finally, the discovery of deserts has become an important mission in the observation and surveillance of Earth by artificial satellites.

Very often, under the pretext of "civilization", safety or "pacification", exploits in the desert would end with the arid regions being seized by foreign powers: the Spanish colonization of the Americas, the Dutch and German colonization of South Africa, the colonization of Australia by the English, Russian colonies in Central Asia and French colonies in the Sahara). The peoples of the desert, scattered and powerless, were never able to effectively resist these takeovers. They were sometimes absorbed into the Continent-States, as was the case in the United States, Australia or the Russian Empire (and later the Soviet Union); in other instances, they were redivided or redistributed under the authority of riparian States. The desert thus became an object of political hegemony and economic exploitation, the subject of rivalries and conflict, and a pawn in geopolitical strategy that went far beyond its borders.

2.11. References

General opus

Aymard, A. and Auboyer J. (1953). "L'Orient et la Grèce antique." In *Histoire Générale des Civilisations*. Ed. by M. Crouzet. Vol. 1. Paris: Presses Universitaires de France.

— (1954). "Rome et son empire." In *Histoire Générale des Civilisations*. Ed. by M. Crouzet. Paris: Presses Universitaires de France.

Durou, J.-M. (1993). *L'exploration du Sahara*. Paris: Actes Sud.

Favre, H. (1997). *Les Incas*. 7th edition. Que sais-je? 1504. Paris: Presses Universitaires de France.

Grousset, R. and Léonard E. (1956). *Histoire universelle*. Gallimard.

— (1957). "De l'Islam à la Réforme." In *Histoire universelle*. Vol. 2. Encyclopédie de la Pléiade. Paris: Gallimard.

Julien, C.A. (1951). *Histoire de l'Afrique du Nord, Tunisie, Algérie, Maroc, des origines à la conquête arabe*. 2nd edition Paris: Payot

Monteil, V. (1964). *L'islam noir*. Paris: Le Seuil.

Perroy, E. et al. (1955). "Le Moyen-Âge." In *Histoire Générale des Civilisations*. Ed. by M. Crouzet. Paris: Presses Universitaires de France.

Puech, H. (1970–1976). *Histoire des religions*. Encyclopédie de la Pléiade. Paris: Gallimard.

Sourdel, D. (1999). *L'islam*. 20th edition. Que sais-je ? 395. Paris: Presses Universitaires de France.

Soustelle, J. (1994). *Les Aztèques*. 7th edition. Que sais-je? 1391. Paris: Presses Universitaires de France.

Stamm, A. (1997). *Histoire de l'Afrique précoloniale*. Que sais-je? 241. Paris: Presses Universitaires de France.

Waldman, C. (1985). *Atlas of the North American Indian*. New York: Facts On File.

Prehistory

Chaline, J. (1996). *L'évolution humaine*. 2nd edition. Que sais-je? Paris: Presses Universitaires de France.

Coppens, Y. (1983). *Le Singe, l'Afrique et l'Homme*. Paris: Fayard.

Picq, P. (1999). *Les origines de l'homme*. Paris: Tallandier.

Picq, P. and Coppens Y. (2001). *Aux origines de l'humanité*. Paris: Fayard.

Pour La Science, ed. (1999). *Les origines de l'humanité*. Dossier hors série. Paris: Belin.

Reichholf, J. (1991). *L'apparition de l'homme et ses rapports avec la nature*. Paris: Flammarion.

Thomas, H. (1994). *L'homme avant l'homme, le scénario des origines*. Découverte 215. Paris: Gallimard.

Antiquity

Amiet, P. (1999). *L'Antiquité orientale*. 6th edition. Que sais-je ? 185. Paris: Presses Universitaires de France.

Briant, P. (1994). *Alexandre le Grand*. 4th edition. Que sais-je? 622. Paris: Presses Universitaires de France.

Ghirschman, R. (1951). *L'Iran des origines à l'Islam*. Paris: Payot.

Grenet, J. (1995). *La Chine ancienne*. 7th edition. Que sais-je? 1113. Paris: Presses Universitaires de France.

Lemaire, A. (1995). *Histoire du peuple hébreu*. 4th edition. Que sais-je? 1898. Paris: Presses Universitaires de France.

Margueron, J.-C. (2003). *Les Mésopotamiens*. Paris: Picard.

Middle Ages

Boorstin, D. (1997). "Les routes de l'Orient." In *Les découvreurs*. Collection Bouquins. Paris: Robert Laffont, pp. 114–141.

Boulnois, L. (1966). *La Route de la soie*. 1987 edition. Paris: Arthaud.

Elisséeff, N. (1979). *L'Orient musulman au Moyen-Âge, 622-1260*. Collection U. Paris: Armand Colin.

Halphen, L. (1948). *Les Barbares, des grandes invasions aux conquêtes turques du XIe siècle*. Paris: Presses Universitaires de France.

Hoang, M. (1988). *Gengis Khan*. Paris: Fayard.

Jaffre, J. and Durou J.-M. (1978). *La caravane du sel*. Paris: Denoël.

Kitsikis, D. (1984). *L'Empire ottoman*. 3rd edition. Que sais-je? 2222. Paris: Presses Universitaires de France.

Lemerle, P. (1998). *Histoire de Byzance*. 13th edition. Que sais-je? 107. Paris: Presses Universitaires de France.

Marçais, G. (1946). *La Berbérie musulmane et l'Orient au Moyen-Âge*. Paris: Aubier.

Mauny, R. (1971). *Les siècles obscurs de l'Afrique noire*. Paris: Fayard.

Morrisson, C. (1998). *Les croisades*. 8th edition. Que sais-je ? 157. Paris: Presses Universitaires de France.

Roux, J.-P. (1982). *Les Barbares*. Paris: Bordas.

— (1984). *Histoire des Turcs*. Paris: Fayard.

— (1991). *Tamerlan*. Paris: Fayard.

— (1993). *Histoire de l'Empire mongol*. Paris: Fayard.

Sourdel, D. (1998). *Histoire des Arabes*. 6th edition. Que sais-je ? 1627. Paris: Presses Universitaires de France.

Travellers and merchants

Heers, J. (1983). *Marco Polo*. Paris: Fayard.

Jan, M. (1992). "Envoyés et missionnaires du Moyen-Âge." In *Le voyage en Asie centrale et au Tibet.* Collection Bouquins. Paris: Robert Laffont, pp. 1–130.
de La Roncière, C. (1925–1927). *La découverte de l'Afrique au Moyen-Âge.* 2 vols. Cairo: Société Royale de Géographie d'Égypte.
T'Serstevens, A. (1959). *Les précurseurs de Marco Polo.* Paris: Arthaud.

Ancient Texts

Al Idrisi (1154). *Géographie.* Trans. by A. Jaubert and A. Net. Paris: Flammarion. 1999.
Herodotus (ca. 443 BC–420 BC). *L'enquête.* Trans. by A. Barguet. Bibliothèque de la Pléiade. Paris: Gallimard. 1979.
Ibn Battuta (1352). *Voyages.* Trans. by C. Defremy and B. Sanguinetti. 3 vols. Paris: François Maspero. 1982.
Ibn Fadlan, Ibn Jubayr, Ibn Battuta, et al. (921). *Voyageurs arabes.* Trans. by P. Charles Dominique. Bibliothèque de la Pléiade. Paris: Gallimard. 1995.
Leo Africanus, H. i. M. a. F. (1556). *Description de l'Afrique.* Paris: Leroux (1896-1898).
de Plan Carpin, J. (1247). *Histoire des Mongols.* Trans. by C. Schmitt. Paris: Éditions franciscaines. 1961.
Polo, M. (1297). *Le devisement du monde — Le livre des merveilles.* Trans. by L. Hambis. 2 vols. Poche 45-46. Paris: La Découverte. 1998.
de Rubrouck, G. (1255). *Voyage dans l'Empire mongol.* Trans. by C. Kappler and R. Kappler. Paris: Payot. 1985.
Strabon (18 AD–24a AD). "Géographie. L'Égypte et l'Éthiopie nilotique." Livre XVII, 1re partie. In Trans. by B. Laudenbach. Vol. XIV. 514. Paris: Les Belles Lettres. 2016.
— (18 AD–24b AD). "Géographie. L'Afrique, de l'Atlantique au golfe de Saloum." Livre XVII, 2e partie. In Trans. by B. Laudenbach. Vol. XV. 504. Paris: Les Belles Lettres. 2014.

3

Aridity

3.1. Where we examine semantics

Right at the start of this book, I highlighted the ambiguity around the word *desert* and the difficulty of using it in a scientific context. Further, in general usage, *desert* is usually associated with two other words: *drought* and *aridity*, which are considered to be equivalent or interchangeable, but which, in reality, are not true synonyms. The semantic confusion that results from this usage is all the more problematic as it hampers the analysis of two distinct phenomena, which have different consequences both on the dynamics of the regions concerned as well as on the life of the people who live there. Let us, therefore, try to develop clarity around these words and how they are used.

A *desert* is an empty or abandoned space, with no life or very little life in it, perceived as having these qualities in comparison to a more populated and attractive environment. The geographic meaning of this word was established over the 18th and 19th Centuries because of explorers, and the deserts that we see on maps today date back to the maps of the 18th and 19th Centuries. However, deserts are far from being homogeneous and they differ in various aspects depending on both time and place. *Desert* or *desertic*, therefore, is only relative in nature, being subjective and vague terms that depend on each person's understanding and experience.

A *drought* is a period in time, whose duration can vary, in which there is deficient water from rainfall. We also speak of a dry spell or a dry season. This is a period with insufficient rainfall that interrupts what is normally a wet season. This results in a lack of water resources that are necessary for the weathering of rocks and for maintaining green cover, and thus brings about degradation in green cover and in

Chapter written by Fernand JOLY.

agricultural practices. Droughts are exacerbated by high temperatures that result in greater evaporation and are, therefore, most dangerous in summer and in hot regions. This condition is especially harsh along the edges of deserts since variations in rainfall are critical in these extreme conditions. As readers may recall, in cold regions and even during winter in temperate regions, water held in the form of snow or ice on the ground may have comparable effects. A drought can be short or very long, and it can be an exceptional occurrence, seasonal or even a multiyear phenomenon. A state of *drought*, as with a *desertic* state, is a relative condition, compared to a state that is considered to be normal and which can be specified and, when possible, quantified.

Aridity is a physical state in nature, which is characterized by a persistent drought. It applies to a place or a region: we speak of arid "lands" or "regions". The term refers to a complex situation that combines several factors that are measurable in theory, but are bewilderingly large in number and are interdependent. In addition to the patent deficiency in rainfall and the excess evaporation, both of which are meteorological conditions, there are local conditions that play a role: biological factors (such as the density of green cover, the rate of transpiration in living beings in the area), topographical factors, (such as the slope systems), hydrological factors (such as types of flow and water storage) and geological factors (e.g. permeability in rocks). A state of aridity is said to be present whenever, over multiple years, the quantity of rainfall does not compensate for the sum of potential water-loss through evaporation, runoff and infiltration:

$$P < ET + R + I, \qquad [3.1]$$

where P is the quantity of water brought in through precipitation, ET is *evapotranspiration* (physical evaporation and transpiration by plants), R is the quantity of water removed as runoff and I is the quantity of water lost through infiltration into rocks, becoming inaccessible to plants or animals.

If the chief cause of aridity is related to the geological structure (stretches of permeable ground such as sand, *karst*[1] or fissured rocks) or the relief features (steep slopes or craggy mountainous regions), then it is said to be *edaphic aridity* (from the Greek edaphos, ἔδαφος, ground). This kind of aridity usually only extends over small or mid-size territories (between 100 and 1,000 km^2): a hillside, a valley, a river basin or a massif.

On the other hand, if the aridity is chiefly caused by meteorological conditions, such as atmospheric circulation, chronic rainfall deficit or lasting effects from evaporation due to heat, it is called *climatic aridity*, which may extend over very large areas (10,000 to 1 million km^2), or even across entire portions of a continent.

1 Karst: limestone relief, where biochemical and geochemical erosion prevails, with underground water flow.

These are the *desert* or *xeric* (from the Greek *xeros*, ξήρος, dry) regions, and it is these deserts that are the main focus of this book.

As can be seen, *desert*, *drought* and *aridity* must not be used interchangeably. A *desert* is, above all, a landscape, an emotional impression that is carried by an observer. A *drought* is a temporary pluviometric phenomenon in an otherwise normally wet climate. Droughts may affect regions that are not arid, such as the temperate regions. They can also cause large-scale damage, as humans are not able to fight or control them. *Aridity* is extreme drought that transcends time and takes over space. Humans either adapt to aridity or abandon arid lands. Aridity is what truly makes a desert.

3.2. Causes of aridity

The large variety in desert landscapes suggests that there are many different causes that play a role in the qualification and location of deserts. Some of these factors are meteorological, with respect to general atmospheric circulation, while others are geographical, such as the region's position with respect to the sea, the relief features or the soil type. There are also anthropic factors that are related to how the land is used by the humans who live here.

3.2.1. *Meteorological causes*

These are the chief causes of aridity, playing a determining role and with the most widespread effects. Meteorological causes are responsible for the largest deserts in the world. Any atmospheric conditions that prevent sufficient rainfall or, on the other hand, facilitate intense evaporation, are potential causes of aridity in regions where they predominate. In general, the processes associated with thermal factors (variations in temperature) and dynamic factors (air movement) determine how much moisture there is in air.

3.2.1.1. *Solar radiation*

The primary factor here is *solar radiation*, the extraterrestrial cause of all energy that is required for the functioning of terrestrial thermodynamic processes. This radiation is directly produced by the Sun's thermonuclear activity and is made up of a spectrum of radiations, ranging from ultraviolet to infrared (IR) wavelengths, with the majority of radiation falling into the visible spectrum of light, between the wavelengths of 0.4 and 0.7 µm. The total quantity of energy that the Earth receives at the level of the atmosphere stays practically the same all through the year and this quantity is, therefore, called the "solar constant" and has a value of

2 calories cm^{-2} min^{-1}. While traveling through the atmosphere, about one-fifth of this radiation (20.5%) is diffused and absorbed by the air, where it is transformed into thermal energy (temperature) and mechanical energy (convection and winds). About half of this radiation (49.9%) reaches the soil directly and sets off chemical reactions (changes in rocks) and biochemical reactions (photosynthesis and metabolism in living beings). The remaining radiation, about 30%, is sent back into space through reflection of clouds and soil. In addition to this overall loss (called *albedo*), there is also the radiation from the ground itself, from the atmosphere and from clouds, which are chiefly produced in the IR spectrum.

The overall equation for the total energy can be expressed in the highly simplified form:

$$e = e_i - e_r, \qquad [3.2]$$

where e is the remaining energy available, e_i is the incident energy reaching the ground and e_r is the total energy reflected back into space. Even in this exaggeratedly simple form, it is difficult to establish and interpret this equation. On the scale of the entire Earth, and over a multiyear period, this equation is practically zero. A positive value would lead to a continuous heating of the planet, while a negative result would lead to the cooling down of the planet. This is likely to have happened in the past, resulting in drastic climate change and it may also happen in the future, especially if humans aggravate the situation through artificial pollution of the air. However, this equation is not the same in all places and at all times. It is modulated by the angle of incidence of the Sun's rays on the soil, which vary depending on the latitude, the season and slopes, as well as on cloud cover and the soil surface, i.e. local geographical conditions.

The positive part of the equation, e_i, is zero during the night and attains its maximum value during the day. Its value increases as the angle of incidence of the solar ray increases (between 0° and 90°), and grows smaller when there are pollutants (clouds, absorbent gases, dust, etc.) that reduce air purity. The negative part, e_r, depends chiefly on the soil's reflectance qualities during the day, and on the IR radiation that the Earth itself emits at night. However, this variable decreases in value if there is cloud cover or gaseous particles, acting like a cover around the Earth, that do not allow the propagation of reflected and IR rays (the "greenhouse effect"). The resulting balance, e, is both fragile and mobile and, directly or indirectly, the other factors determining aridity depend on this.

3.2.1.2. *Humidity in air*

Absolute humidity is the quantity of water vapor in the atmosphere at a given point in time. This depends directly on temperature and the availability of water from evaporating surfaces such as the sea, lakes or rivers, and from plants (evapotranspiration). A lack of such resources means that in the deserts, water vapor in the atmosphere comes chiefly from the exterior, through cyclones, monsoons or carried in on the sea breeze. Thus, absolute humidity is much greater in the Southern Hemisphere than in the Northern Hemisphere, as more of the oceans lie in the Southern Hemisphere and, similarly, coastal deserts have a higher level of humidity than continental deserts. At a given temperature, absolute humidity has a maximum value that increases as temperature increases. When this limit is attained, the air is said to be saturated. Therefore, the atmosphere in hot deserts is theoretically capable of holding more water vapor than the atmosphere in cold deserts. However, it is very rare that the atmosphere in these deserts does, in fact, reach saturation.

Relative humidity is the ratio between the quantity of water vapor that is actually present in the atmosphere and the quantity that it would hold at saturation. By deduction, this ratio, or hygrometric degree, indicates the amount of water vapor that the atmosphere could absorb and this value increases as the hygrometric degree decreases, i.e. the smaller the ratio, the greater this potential value. Relative humidity is generally very low in deserts, especially in continental deserts: less than 60% or even 40% in the summer in temperate or cold deserts, and less than 50% or even as low as 20% in hot deserts. In the central and Eastern Saharan region, hygrometric degrees in different months have measured 40–50% in winter, while in summer the measurement has been 15–20%. These measurements could even go below 10%, as low as 2–3%. This indicates the potential of absorption of water vapor into the air, and consequently, the actual drying power of the atmosphere in these deserts. On the other hand, relative humidity is higher in the deserts of southern Africa and Australia due to the presence of oceanic bodies. This value can reach its maximum at 90% or even 100% in coastal tropical deserts such as the Namib or Peru.

When the temperature of a saturated, humid atmosphere decreases, the water vapor that it contains condenses and forms clouds. Any cause that results in a decrease in the temperature of a mass of air is a possible cause of condensation, for instance the mass of air rising upwards, with thinner air at higher altitudes, or when a mass of warm air comes into contact with a mass of colder air in general atmospheric circulation. Nebulosity, expressed on a scale of 0–10, is the proportion of cloud-covered sky in the total area of the observed sky. It is greater in winter than in summer, greater in cold deserts than in hot deserts, greater along the coast than inland, and, for the same latitude, greater in the humid atmosphere of the Southern Hemisphere than in the Northern Hemisphere. Clouds are made up of fine particles of liquid water (solid if the temperature is below 0°C). When clouds attain a sufficient mass, these water particles fall to earth as precipitation in the form of rainfall, snow or hail.

When there is a drop in temperature close to the ground, it produces mists or fog. On the ground itself, on a cold surface or wall, or one that has cooled down due to nocturnal IR radiation, water vapor condenses directly on rocks or vegetation in the form of dew or frost (around 0°C). This is what we call "occult precipitation", as it is hard to measure. In certain cases, however, these occult precipitations are essential sources of water: in the Namib, for example, where there is 40–50 mm of dew, as against only 15–20 mm of rain, as also in Peru and in Israel, among other regions.

3.2.1.3. *Rising air and subsidence*

Hot air or warmed air is light and tends to rise up, thus creating a barometric deficit, or low pressure zone, near the ground. The air that rises moves from a low and warm region (in absolute or relative terms) of the atmosphere to a region at a higher altitude, that is cooler (adiabatic gradient) and less dense. The air becomes thinner and cools down, its relative humidity increases and the water vapor that it contains condenses into clouds (nebulosity) or rain (precipitations). Any cause for the rising up of air, whether thermal (nebulosity) or dynamic (air traveling over relief features, or the rise of hot air over a cooler mass of air), is a possible cause of precipitation. Similar phenomena are produced during the horizontal displacement of a mass of humid and cool air over a region with higher temperatures.

Cold air, on the other hand, is heavy, sinks down and accumulates close to the ground, creating high-pressure areas. In subsiding air currents, the air that is descending gets warmer. This is partly because it is moving from a cold region to a warmer region, and also because it shrinks and gets compressed (adiabatic compression). Relative humidity decreases, moving further away from saturation point and this allows a large absorption of water vapor. Nebulosity disappears, thus decreasing the chances of precipitation, and evaporation increases. Subsiding air is drying air.

3.2.1.4. *Hadley cells*

Aerodynamic conditions of this kind are found around the world on either side of the equator, between latitudes 20° and 40° North and South, and are a result of tropical atmospheric circulation.

The conventional explanation for this is derived from a thermal process proposed by J. Hadley in 1735. Above the equatorial zone, which is constantly warmed by the Sun being close to the zenith, the warm air rises, creating an in-draft near the ground, like the draft created at the bottom of a chimney. This causes the trade winds from the North and South to converge – these are constant winds that were the driving force behind old sail-boats. Drifting with the rotation of the Earth, the trade winds blow from NE to SW in the Northern Hemisphere and from SE to NW in the

Southern Hemisphere. As it rises above the equator, the humid air brought in by the trade winds becomes more dispersed and cools down, which produces high nebulosity (the doldrums), accompanied by heavy precipitation. At higher altitudes, the dry and cooled air moves toward the poles in the opposite direction to the trade winds and again moves down toward the ground. Part of these winds reaches as far as the middle latitudes, where it comes into play in atmospheric circulation around the temperate zone. Another part of this wind gathers near the ground in the tropics, creating, at low altitudes around 30°N and S, high-pressure areas (tropical anticyclones), which are very dry and which form the large deserts in the hot regions.

While Hadley's theory was not entirely wrong, certain aspects had to be corrected (Leroux 1996). In reality, subtropical anticyclones do not form as a consistent and continuous high-pressure belt on either side of the equator. Instead, they are distributed in anticyclonic agglutinations (AA) that are distinct and deformable, organized based on the flow of air masses, the divisions between continents and oceans, and the relief configurations in a region. Unlike in Hadley's theory, these anticyclones are not caused only by subsiding flows from counter-trade winds, which are too weak to be able to cause such high pressures by themselves. On the polar side, they are also caused, and chiefly maintained, by cool and dense air carried through mobile air droplets (*mobile polar anticyclones* (MPA)) from high thermal pressure zones in the Arctic and Antarctic. Winds rotate around each cell (AA) in a clockwise direction in the Northern Hemisphere and anti-clockwise direction in the Southern Hemisphere. On the eastern face of the cells, the NE winds in the north and the SE winds in the south form trade winds, regular winds that grow warmer and that are especially drying, which are juxtaposed with a flow from the east on the equatorial side. Each cell thus determines an area that is arid, even over the oceans, resulting both from tropical subsidence as well as the insufficient saturation of the trade winds. Around the equator, the trade winds converge (intertropical convergence) and thus bring into contact, without mixing however, air from the Northern Hemisphere and the Southern Hemisphere, all along a discontinuous area called the *meteorological equator* whose path along the ground shifts and reforms based on the apparent movement of the Sun.

3.2.1.5. *Seasonal variations in the intertropical system*

The seasonal shifting of the meteorological equator results in a general shift in the latitude of the intertropical climate system as a whole. Between one season and the next, one of the meteorological hemispheres encroaches upon the other, leading to noticeable dynamic and climatological changes over the entire region of tropical arid lands.

In the Northern Hemisphere, this phenomenon is particularly remarkable over the meridian lines that include the western part of the African continent and the eastern Atlantic Ocean.

During the boreal winter (December, January, February), the meteorological equator and the pluvial period that accompanies it are shifted by 10–15° toward the south, and the Saharan anticyclone (AA) is also shifted by the same degree. The MPA that are cold, dense and very active are separated by the low-pressure corridors that run along the northern edge of the meteorological equator from Spain to Egypt and to Arabia, along the Mediterranean, as far as the tropics. The passage of these low-pressure regions can cause a sharp drop in temperature and rainfall (or sandstorms) along the entire northern Saharan region and a strengthening of the trade winds on the eastern edge of the anticyclone. This continental trade wind (the *harmattan*[2]), which is hot and very dry, sweeps across the central and southern Sahara, where there are no relief features apart from the Hoggar and Tibesti mountains. On its western edge, the Saharan anticyclone comes up against the oceanic anticyclone (AA), called the Azores anticyclone. Other MPAs flow in the space between these two anticyclones, approaching the continent from southern Morocco and Mauritania. They feed the maritime trade winds that blow toward the SW, leading to arid conditions along the length of the coast, from Cape Juby to Senegal.

During the boreal summer (July, August, September), there is a decrease in the MPAs and they are less active, remaining largely restricted to the eastern Mediterranean. The winds that escape this region toward the SW (etesian winds) become a hot and dry harmattan over the African continent and over the entire septentrional Sahara. The Saharan anticyclone, driven north by the rise of the meteorological equator and receiving insufficient cool air due to the lack of polar air, weakens under high temperatures. It is even locally replaced by thermal depressions on the Tanezrouft, in northern Sudan, and southern Arabia. These depressions draw in a trans-equatorial tropical flow of wind (the monsoon) formed by the southerlies that drift toward the NE after crossing over the geographic equator. The monsoon carries in the more humid marine air and rain over the southern Saharan region and the Sahel, from Gambia to Chad, as far as Ethiopia. On the other hand, over the Atlantic Ocean, the Azores anticyclone stretches as far as Europe and continues to maintain the maritime trade winds and maintains arid conditions over the coast of Mauritania.

In tropical North America, this mechanism is disrupted by the fact that at the same latitudes the continent is narrower than Africa and that all along the western coast of the continent there are mountainous ranges along the meridian which often exceed

2 Harmattan: a very hot and dry wind that blows from east to west in West Africa.

3,000 m in altitude. The MPA originating in the Arctic thermal anticyclone circulate on either side of this mountain chain. On the Pacific Ocean, in winter and in summer, on the eastern side of the powerful subtropical anticyclone (AA), called the Hawaii anticyclone, the MPAs feed a maritime trade wind, which creates an arid situation along the Californian coast that is comparable to the conditions along the coast of Mauritania. The eastern slopes of the mountain chain lie in the Atlantic meteorological space. In winter, the MPAs, flowing from north to south (or the *northers*) along the central plains, reach the tropics. Dry and cold, they block the penetration of the humid Atlantic air in the southern United States. In summer, the meteorological equator shifts up almost as far as the 20th parallel, while the MPAs are less powerful and do not go past the 35th parallel to the south. However, with the significant thermal contrasts between the cold polar air and the hot tropical air, violent tornadoes are formed toward the end of spring, posing a threat to northern Texas and the entire Midwest region of the United States. The Atlantic air is able to penetrate the open spaces lying between the Rocky Mountains and the Gulf of Mexico, as well as along the low heated plains, bringing with it beneficial rainfall. Nonetheless, in the meteorological account we saw, the distribution and definition of subtropical North American deserts are chiefly dependent on the relief features.

3.2.1.6. *Atmospheric circulation in the Southern Hemisphere*

In the Southern Hemisphere (Hobbs, Lindesay, and Bridgman 1998), atmospheric circulation is governed by the dominant presence of the oceans (81% of the area in this hemisphere) by the presence of the enormous Antarctic ice sheet (12–14 million km^2) and, with the exception of Australia, by the tapering away of continents toward the south. With less stark thermal differences and the absence of any major relief feature apart from the Andes mountains, the circulation here is much more regular than in the Northern Hemisphere. Escaping the constant thermal anticyclone located over the Antarctic continent, the mobile polar highs (MPAs) can travel without any obstruction over the Southern Ocean, reaching as far as the 40th parallel. From here, the MPAs penetrate more or less deeply into the north of each of the three large, connected ocean basins: the Atlantic, Pacific and Indian Oceans.

The MPAs are most active during the austral winter, which corresponds to the boreal summer (July, August, September), stimulated by the drop in temperature. In the Atlantic Ocean, they advance toward the north, almost as far as the equator, going along the African coast. On the eastern side of the oceanic anticyclone (AA), called the Saint Helen anticyclone, these MPAs reinforce the SE-NW maritime trade winds and the aridity along the Namibian coast. The maritime trade winds even cross the geographical equator in order to become the African monsoon, pulled in by the Saharan depressions. On the northern side of the anticyclone, the trade winds feed a flow from the east, which blows over all of Northeast Brazil. Another MPA trajectory reaches the southern top of South America, dividing into two parts here, one on either side of the Andes chain. The MPAs that circulate in the east, over the eastern piedmont, feed the strong winds from the south and southeast that blow over Patagonia and maintain cold and dry conditions in the Argentinian plains and the Chaco region.

Contrarily, during the Austral summer (December, January, February), the MPAs and the Saint Helen anticyclone lose vigor and retreat to the south, though this is a modest retreat, only between 10°–15° in latitude, since the difference in temperature between summer and winter is much smaller here than in the Northern Hemisphere, given the vast expanse of oceans. As a result, the summer circulation is quite similar to the winter circulation. The trade winds, though weaker, still maintain the extreme aridity along the coast of Namibia. The only difference is that the African monsoon does not reach as far into the continent and no longer reaches the Sahara. On the northern side of the Saint Helen anticyclone, the subequatorial eastern flow that reaches the South American continents curves toward the SW and sweeps across the entire coast of Brazil, as far as São-Paulo.

Meteorological conditions in the Pacific Ocean are characterized by the existence of a powerful subtropical oceanic anticyclone off the coast of Chile, called the Easter Island anticyclone. During the austral winter, this anticyclone is fed by cold air and reinforced on its eastern front by the MPAs that flow along the barrier formed by the Andes. The maritime SSE-NNW trade winds, which are well fed, maintain aridity along the coasts of Chile and Peru. Along the northern front of the anticyclone, which is more or less divided into secondary anticyclonic cells across a space that is chiefly maritime, there circulates a flow from the west, pulled toward the NW and then, after it crosses the equator, moving toward the NE, where it forms the Chinese summer monsoon. During the austral summer, the Easter Island anticyclone, as well as the MPA trajectories, retreat toward the south and the trade winds are weakened. Along the American coast, however, the winds remain powerful enough to maintain conditions of aridity along the shores of Chile and Peru. In eastern Australia, the subequatorial flow from the east comes up against the orographic barrier of the coastal ranges, the Great Dividing Mountains, without being able to cross over them.

3.2.1.7. *The unique case of the Indian Ocean*

Unique meteorological conditions are created between Africa, India and Australia by the vast Indian Ocean, a primarily austral ocean, coming into contact with the massive Asiatic continent, which is entirely boreal. These conditions are strikingly contrasting over time, diversified across space and have led to the formidable reputation of the Indian monsoon, which goes back centuries.

During the boreal summer/austral winter (July, August, September), the maritime trade winds originating in the tropical high pressure zones of the Southern Hemisphere (AA) fed by the powerful MPAs from the Antarctic, blow toward the NW. Because of the sharp northward shift of the meteorological equator, these trade winds cross the geographic equator and are pulled in by the thermal low pressure zones that are set up over North India and transform into the SW-NE monsoons on the western coast of the peninsula (the Indian monsoon) until they are blocked by the insurmountable barrier formed by the Himalayas. In the course of its passage, the western flank of this SW monsoon touches the eastern coast of Africa, north of the equator, and drawn in by the thermal low-pressure zones in Sudan and southern Arabia; it penetrates into Somalia and as far as Ethiopia, where it produces only scanty rainfall. Passing across the Gulf of Oman does not give it any additional humidity and the little humidity these winds contain comes down as rainfall over the mountains in southern Pakistan and over the Indus plain (Sindh) and, consequently, when the winds reach the Thar Desert they are dry. To the south of the equator, the easterly flow of the trade wind that flows over the septentrional edge of the oceanic subtropical anticyclone called the Mascarene anticyclone, hits the eastern coast of Madagascar and Mozambique with full force. At the same time, over southern and central Australia, MPA droplets flow from west to east, as far as the Great Dividing Mountains. Between each of these streams of droplets there is a low-pressure corridor set up, with the possibility of rain and the winds blow around them in an anti-clockwise direction. At the head of the flow, maritime and flowing S-N, they are cold and humid winds, sometimes accompanied by tornadoes. The northern front of these winds feeds the SE trade winds that flow over the entire northwest region of Australia. The tail of the winds, continental and oriented N-S, are hot and dry and often carry sand or dust.

During the Austral summer (December, January, February), the entire system weakens and re-descends to the south, following the retreat of the meteorological equator. The boreal trade winds, NE-SW, which gain force from the fixed or mobile high-pressure zones over the Aral-Caspian region of Turkestan, flow along the eastern coast of Africa, becoming dry, continental winds over the Horn of Africa and over Kenya. Crossing the geographical equator, it becomes the NW-SE Malagasy monsoon, bringing rainfall over northawest Madagascar and northern Mozambique. The eastern shores of the Indian Ocean are ruled by the monsoon originating from the high-pressure zones and the winter MPAs from East Asia and the North Pacific. When this monsoon, in turn, crosses the geographic equator, it becomes the NW-SE Australian monsoon, sweeping across the whole of northern Australia, as far as the 20th parallel. Relegated further south in all regions, the austral trade winds still persist in maintaining dry conditions over southern and central Australia, southwest Madagascar and the Kalahari basin.

3.2.2. *Geographic causes*

The general circulation in the atmosphere and meteorological vicissitudes are not the only causes for the distribution of deserts around the world. There are also geographic factors that may intervene to cause aridity or exacerbate existing arid conditions in a region, either by creating obstacles that prevent the penetration of humid winds, or by creating local conditions that are not conducive to the presence of moisture in the air or on the ground.

3.2.2.1. *Continentality*

Continentality is a measure of the distance from the sea. In order to be a cause for aridity, the continentality must be to a degree where the humidity carried by the air reduces progressively as it passes over the continent. The movement of a mass of air over a hot continent (in summer, or in the lower latitudes) increases the temperature of the air mass and, therefore, its evaporation capacity. In winter, and over cold continents, thermal high-pressure zones hinder the penetration and condensation of humid winds. Thus, local atmospheric conditions may in themselves be of a drying nature and there is greater arid degradation when thermal contrasts are more pronounced.

These are the conditions found across the whole section of the Eurasian continent lying in the middle latitudes of the Northern Hemisphere. In winter, thermal high-pressure zones predominate over the cold land in Central Asia and are reinforced by the active flows of polar air from the boreal ice sheets. These flows are blocked in the south and east by mountain ranges that stretch from the Caucasus to North India, China and up to Mongolia. As a result, these MPAs accumulate into an air mass that is dense, very cold and very dry, which becomes stagnant in a continental AA cell that is very stable and covers all of Central Asia. In summer, the circulation of the MPAs weakens and barometric pressure reduces as the regions grow warmer while still remaining higher than barometric pressures in regions that are further south. Thermal depressions now form over the Gulf of Oman and over North India. The flow of the Indian monsoon, which is drawn in by these regions, cannot cross the Himalayas and, therefore, this monsoon does not touch Central Asia. Throughout the region, the barometric conditions and the relief block the penetration of humid, maritime air and aridity reigns here in all seasons. At the heart of the continent, winter is very cold and the soil freezes, adding edaphic aridity to the climatic aridity. This winter is in stark contrast to the very hot summer, where, in the absence of the monsoon, rainfall is chiefly due to the rise of local storms. While these are the overall conditions that prevail, there are, however, certain subtle differences between the western and eastern parts of the continent.

In the west, the continental anticyclone in winter lies largely over the Turan Depression (Aral Sea) and frequently extends as far as Eastern Europe, north of the Black Sea and the Iranian Plateau. This anticyclone blocks or weakens the migration of the MPAs that descend on either side of the Ural Mountains, while it feeds the NE trade winds flowing toward Arabia and the Horn of Africa. This NE trade wind crosses the equator and becomes the Malagasy monsoon. Glacial aridity then sets in from Lake Balkhash to the Caspian Sea and in the Iranian basins of the Great Salt Desert and in Lut. In the summer, this entire region, which is inaccessible to the Indian monsoon due to the mountains in southern Iran and Afghanistan, remains dry. The rare masses of humid, maritime air from the Atlantic that reach this far have lost their humidity en route and, as they travel over the overheated lands, they can become more and more drying as their evaporating capacity increases. An area of permanent aridity thus extends across all of Turkestan, southern Russia as far as the Volga, and a large part of Iran and Afghanistan.

Toward the east, in East Asia, the continental anticyclone is fed in winter by Siberian MPAs, which pass through corridors in Dzungaria and the Gobi Desert, as well as others that go up the course of the Lena, and then continue along the Pacific coast up to the tropical latitudes. In summer, the aridity there is less pronounced than in the West. This is because, although the Indian monsoon remains blocked by the Himalayas, there is the advection of humid, maritime air (the Chinese monsoon) that can come in from the Pacific Ocean through North China and Manchuria, and reach the eastern regions of the Gobi Desert and Siberia.

This kind of an arid continental system is unique and does not exist anywhere else in the world on such a large scale and at this latitude. In temperate North America, where there is a high degree of continentality in the large plains that lie along the main trajectory of the north-Canadian MPAs, the winters are harsh and generally dry, as maritime air from the Atlantic rarely enters there or is blocked by the Appalachian Mountains. In summer, on the other hand, maritime air flows are quite frequent, coming in from the Gulf of Mexico. True desertic aridity, therefore, is never established anywhere in this region. Continentality does, indeed, play a role in other parts of the world in the tropical latitudes: in the Sahara, the American West, southern Africa and Australia. However, its role here is secondary to other factors that play a greater role in establishing aridity, namely meteorological conditions and relief features.

3.2.2.2. *The unique features of ocean margins*

At first sight, it would seem paradoxical to note that the deserts located along the coast are among the driest of all deserts. This is, however, the case with all deserts located on the western coasts of the tropical continents and these offer a very good illustration of the complex and highly interdependent relationships that exist between atmospheric phenomena and oceanic circulation around the shores.

Why does this happen only along the western coasts? This is because the eastern coasts of continents (at the level of the latitudes under consideration) are constantly subject to easterly oceanic winds, the maritime trade winds in the winter and the humid monsoon in the summer. There is no real aridity along the coast here (except in Arabia and Somalia, which are constantly exposed to the dry, continentalized winds), even if aridity can be observed a little further inland. On the western edges of the continents, however, the prevailing winds on the coast are drying trade winds, blowing on the eastern front of the large subtropical ocean anticyclones (AA). This trade wind, blowing from the NE or the SE, along a direction that is more or less divergent with respect to the coast, tends to push the layer of warm water on the surface away from the coast (over several dozens of kilometers and up to 100–200 m of thickness, more or less). The relative void that is thus created is then filled in by the upwelling of cold water from the depths, which, when dragged away in turn by the trade winds, forms "cold currents" along the coasts: from the Canary islands in the North Atlantic along the coasts of Morocco and Mauritania, from Benguela in the southern Atlantic, along the coasts of Namibia and Angola, in California (from San Francisco) up to Mexico in the North Pacific, from Humboldt to Chile and to Peru in the South Pacific. This phenomenon may also occur along the western coast of Australia, but is rarer and occurs, if at all, only in winter, when the trade winds are maintained by the frequent passage of the mobile Antarctic anticyclones.

This state of things produces a highly negative difference in the temperature of waters close to the coast, between 5°C and 10°C lower than the waters further out to sea. When coming into contact with this cold water, the temperature of the lower layers of the atmosphere also decreases, which makes them more stable and, therefore, reduces the chance of a thermal upwards movement of pluviogenic systems. At the same time, this fall in temperature increases the hygrometric degree up to the point of saturation with the formation of mists and dew. There is high nebulosity but no rainfall: coastal deserts are "fog" deserts, with fog extending inland up to various degrees, depending on the proximity of steep relief features on the coast.

These stable, well-established systems may be disrupted by temporary anomalies in oceanic circulation. The most important of these that concerns coastal deserts is the phenomenon called the *El Niño* (the Infant Jesus) by the Spanish colonizers of Peru. Peruvian fisherman had long been aware that in certain years, over the Christmas period, the coastal waters would become unusually warm. They noticed that at the same time there was a near-total disappearance of anchovies, their staple catch, and of birds, which produce guano. This was resulting in economic crises for the fishermen, but these conditions were beneficial to neighboring farmers as they produced exceptional rains. However, these rains were also always accompanied by the risk of floods and catastrophic landslides. It was only in the 20th Century that El Niño was recognized as a recurrent phenomenon, occurring every 4–7 years and whose effects were much more far-reaching than a simple change in local weather patterns (Voituriez and Jacques 1999).

El Niño is a major oceanic-atmospheric equatorial event, which is particularly strong in the Pacific Ocean. It could only be understood following satellite observations of simultaneous variations in sea levels and water temperatures. In a normal situation, while in the Americas the trade winds cause the upwelling of "cold" waters (20–23°C) to the surface and provoke desert conditions on the mainland, the easterlies and southern and northern equatorial currents cause warm water (28–29°C) to accumulate in the west, which is conducive to evaporation and rain over northeastern Australia and eastern New Guinea. However, in certain years, this western reservoir of warm water starts to shift toward the east in spring (March–April). Strengthening the W-E equatorial countercurrent, this water reaches the American coasts by the end of summer and in November–December, spreads out to the maximum along the Californian coast to the north and the Peruvian coast to the south. With the arrival of this warm water, atmospheric pressure reduces; the trade winds weaken, the upwelling disappears and there is the onset of rain over the western parts of the Americas, from the coast of Oregon down to Chile, with the desert sprouting green cover. However, these conditions produce a severe drought, which can lead to massive forest fires in Australia and Indonesia (Siegert *et al.* 2001.

The El Niño event is therefore linked to this movement of warm water, which is itself associated with the periodic shifts of subtropical high-pressure regions across the southern Pacific, a phenomenon called El Niño–Southern Oscillation (ENSO).

The root causes for this phenomenon are still not clear, despite the many resources now available to oceanographers and meteorologists. Do sunspots play a role? Is there a periodic overspill from the dynamic boreal atmosphere? Or is it because of the notorious greenhouse effect and global warming? It is common, today, to attribute all kinds of climate disasters around the world to the El Niño event. And in fact, it is now even paired with the La Niña effect, which is the converse phenomenon (i.e. a return to normalcy) that brings back drought, forest fires and a loss of green cover and fauna in the Americas, while bringing rain and floods to Australia, Indonesia and Southeast Asia. At any rate, it has been observed that the El Niño events have become increasingly frequent (becoming almost an annual phenomenon since 1990), that they grow more intense and that the interval between two destructive Niños is shrinking. It is also observed that in the tropics, on either side of the Pacific, El Niños and La Niñas cause thousands of deaths and damages worth billions of dollars among poor populations and vulnerable economies.

Do El Niños also exist in the Indian and Atlantic Oceans? There is no definite answer yet. Statistically, certain pluviometric phenomena over these oceans do seem to synchronize with strong El Niños or La Niñas. For example, in 1997–1998, (*idem* late 2004) a strong El Niño ("El Niño", the "El Niño of the century") coincided with heavy rains in Central America and in the coastal Andes as well as in East Africa

(Somalia, Kenya, Tanzania), while there was severe drought in Southeast Asia and Australia, as well as in the upper regions of the Central Andes, Northeast Brazil and West Africa (the Sahel and the Namib Deserts). Conversely, during summer in the year 2000, as well as in June 2002, massive fires ravaged the forests in 11 States in the American West, from Idaho to California and New Mexico, while Southeast Asia faced floods, Northeast Brazil suffered torrential rains and landslides, and Western Europe saw a wet summer. An increase in the frequency of these events was even cited as the cause for the duration and intensity of certain Sahelian droughts, like in 1910–1915, 1940–1949, 1968–1973, 1977–1985 and 1997–1998. It remains to be seen if there is indeed a correlation between these events or whether they were the result of a combination of common causes involving global atmospheric and oceanic circulation. Whatever the reason, the fact remains that the arid coastal climates in tropical regions are significantly affected.

3.2.2.3. *The dynamic role of relief features*

Topographical relief features play various roles in modifying dynamic, thermal and hygrometric conditions of air currents, mainly through their altitude and mass, and also by their orientation.

Relief features across the Earth's surface reach altitudes of less than 9000 m – the highest surface being that of Mount Everest, in the Himalayas, reaching a height of 8,846 m. The entire surface of the Earth is, therefore, covered by the troposphere[3] and, over the majority of Earth's surface, by the turbulent layer, which is most disturbed between 0 and 3,000 m. Further, as altitude increases, there is a decrease in pressure (the barometric gradient) and temperature (the adiabatic gradient): the air becomes rarefied, colder and less heavy. At the same time, the total quantity of water in the air (absolute humidity) reduces: 80% of water vapor in the atmosphere lies below 4000 m. Beyond a certain altitude, marked by a maximum of precipitation, the chances of rainfall in dry air grow increasingly slimmer. This sharp hygrometric and pluviometric fall leads to aridity at high altitudes, which is a characteristic of high mountains.

The shift in the bioclimatic situation from a piedmont to aridity at high altitudes takes place in gradual stages, localized within a regional climatic pattern. The variations in these stages are chiefly the values of each parameter (rainfall, temperature, density and the composition of the vegetation) rather than the distribution of these parameters in time (the pattern): winter rains occur in winter and

---

3 Troposphere: the lowest layer of the atmosphere, the site of atmospheric turbulence, located below the stratosphere.

summer rains in summer; in the tropics, the length of the summer day remains shorter than in the temperate regions, while the winter days remain longer than in the temperate regions, and, finally, the verticality of the Sun's rays are more marked here than anywhere else on Earth. The other things that change, however, are that with lower temperatures at higher altitudes, precipitation occurs either partially or entirely in the form of snow, which may accumulate locally to form glaciers. Pluvial aridity is compounded by snow retention and the edaphic aridity of frozen soil. The lower limit for permanent snows is higher in the tropics (5,500 m in the Peruvian Andes) than around the middle latitudes (3,900 m in the Sierra Nevada in California, and 2,300 m in the Altaï). The lower limit for aridity caused by altitude is higher in humid climates than in dry climates. And if the piedmont is in itself desertic, then the entire mountain risks being desertic.

However, altitude is not the sole cause for aridity. We must also take into consideration the role played by relief, depending on whether it is a largely regular massif or, on the contrary, if it is articulated and highly differentiated. In the first case, aridity then tends to be distributed evenly across the region as is the case, for instance, in the Sahara, in Turkestan, in southern Africa or in Australia. The differences in relief here are limited to those that are caused by the latitude and by regional mesotopography. In the second case, as in the Mongol or Chinese regions of what used to be known as "Upper Asia", or across the highlands in the western United States, the arranges of ridges, chains and valleys intensify the effects of the contrasts, which are mostly linked to the orientation of the orographic system with respect to the Sun's rays or to air flow.

When the Sun's rays fall upon a slope, the slope is warmed to a greater degree when the angle of incidence between the Sun's rays and the slope is closer to 90°, and when the air at higher altitudes is more rarefied and drier. However, the slope that is not exposed to the Sun's rays is at a disadvantage, colder and lies more or less in shadow. At night, the opposite effect is produced: cooling due to the earth's radiation is stronger when the diurnal heat received through sunlight was more intense, leading to a greater thermal difference along the slopes that are exposed to sunlight than the shaded slopes. These contrasts, which can be seen in the vegetation and dynamism along the slopes, have long been noticed in mountains in temperate zones. The phenomena are even given specific names in the Alps: "adret" for the warm slope, and "ubac" for the cold slope. In deserts, one is arid, while the other is merely dry. These contrasts are sharper in the middle latitudes (where the Sun's rays are more slanting, forming deeper shadows) than in the intertropical latitudes, where the angle of incidence can even be 90°. The differences are also more marked in mountain chains oriented E-W, running crosswise with respect to the Sun's rays (as in Asia) than in mountains oriented along the N-S axis, which receive sunlight longitudinally (as in America). These factors explain why mountains in cold or temperate deserts usually offer a larger variety of types of aridity than mountains in hot deserts.

An even more important factor, with respect to aridity, is the presence of an orographic barrier that stands perpendicular to the flow of saturated air. When the flow comes up against the side facing the wind, it is forced to rise, with the air growing less dense and the temperature falling. The humidity in the air then condenses, forming a band of clouds, with or without precipitation. The heat released by this condensation can, to some extent, slow the cooling (moist adiabatic lapse rate: 0,5°C per 100 m), but the air reaches higher altitudes, having lost a considerable amount of moisture and with a temperature that is significantly lower than the air at the base of the slope. On the leeward side, the air descends once more, gets compressed and its temperature rises much faster than it had decreased during the ascent (dry adiabatic lapse rate: 1°C per 100 m). At the same altitude, the temperature on the leeward side of the relief may be over 10°C higher than the windward slope. Not only do clouds not form here, but the air becomes particularly drying: if there is any snow, it melts; fires can start in the vegetation and evaporation removes all humidity from the ground. This is called the "Foehn effect", named after the wind produced in the Alps when saturated air from the south, having led to abundant precipitation on the Italian slopes, comes down as hot and dry air on the North Tirolean slopes.

Similar effects, often known by different local names, can be found wherever a high relief feature comes up against a strong and very humid air current: in Asia, while the Indian monsoon is blocked by the Himalayas, which are too high and too rigid for it to cross, it does cross over mountains in Afghanistan, further to the West, and blows as a dry and hot wind, called the "Afghanets" over the plains of Turkestan; in the United States, the Native American word "chinook" is used for the westerly wind from the Pacific that crosses the barrier of the Rocky Mountains, coming down again as a drying wind over the plains of Missouri; and, in Argentina, the hot, dry wind blowing from the eastern slopes of the Andes is called the "zonda". Orographic ascent and subsidence create large climatic asymmetries (the scale of which depends on the intensity of the flow and the strength of the obstacle), which are the cause for a specific type of desert that often cover a small area but are always very harsh, called *rain shadows*. A final aspect of the role played by relief is the non-negligible contribution of topography. Steep slopes accelerate surface flows and, in equal measure, reduce the chances of infiltration. They thereby amplify the causes for water loss from the initial pluviometric catchment and, therefore, increase the causes for aridity. With precipitation being equal, an uneven water catchment area is statistically more arid than a flat catchment area.

3.2.2.4. *Edaphic aridity*

Aridity, which is a lack of water, can also manifest itself as soil aridity or *edaphic aridity* (without surface water: Monod), which is caused by the infiltration of water that has fallen, or by runoff due to the soil permeability. If the permeable formation is very thick, the absorption of water may be total and the surface may be completely deprived of water. This total disappearance from the surface may lead to the water becoming unavailable to plants and animals. When this happens, soil aridity is then biologically expressed by very poor, or even a complete lack of vegetation and fauna. This aridity is essentially caused by the local geopedology and may occur even in a humid climate (e.g. Causses, in southern France, Yucatán in Mexico). How the water that comes from the atmosphere (and which is needed by all living things) is stored is, therefore, an important factor (Volume 2, Chapter 1), which is itself governed by various criteria.

The first of these are lithological structures and their fracturations. This relates to the problem of *permeability*. In coherent rocks, fractured by tectonic activity, water enters even more easily and penetrates deeper when fissures are wide and interconnected. If the rock is soluble, as is the case with limestone, excess water causes the fissures to widen and combine, creating a veritable underground drainage systems – this is called the karst phenomenon. If the formation is very thick and the karst is deep, there may be a complete absorption of water and the surface may be completely deprived of water. As mentioned before, this aridity is essentially caused by geology and may occur even in humid climates. In loose rocks, the penetration (and thus loss) of water depends on the dimension and the connectivity of voids (pores) between the grains. This penetration is rapid and fleeting in rocks made up of large components (conglomerations of blocks or gravel) and is slower and more accumulative in porous rocks (fine sand or silt), with almost no penetration into clays. Thus, aridity is always more marked in regions where the first type exist than in the presence of the other type of rocks.

Another criterion, equally important, is the presence or absence of detritic surface formations, which may be recent or have accumulated over the years. This presents the problem of *infiltrability*, which is the aptitude of water transfer between the surface and the rock. The surface layer acts as a buffer layer, an intermediate layer whose properties can either facilitate or block the flow of water deeper into the soil. When there is no such cover, or if it is only a thin layer, the penetration depends on the rock itself, but the water has a greater tendency to trickle away, especially if there is low fissuring, high pluviometric concentration and sparse vegetation. However, a thick layer of non-clayey soil (apart from crusts or other impermeable formations) and denser vegetation, especially if it is herbaceous, are more conducive to infiltration. A special case, commonly observed in cold deserts, is when water on the ground is frozen, either seasonally or permanently (permafrost). The ice here blocks interstices and makes the soil impermeable as long as the ground remains frozen.

Thus, after climatic conditions, it is aspects of surface, relief, soil and vegetation that chiefly govern the existence or the continuance of aridity. These stable factors regulate hydrological reserves and microclimates. Modifications in these factors create disruptions that are more or less reversible depending on the situation. For example, wind or water erosion of soil denudes rocks and leads to a decrease in infiltration and scarcer vegetation, which could produce a temporary drought or even aridity. However, these effects are always limited to a region and create small subsets of relatively restricted conditions within a larger set of arid climatic conditions.

3.2.3. *Human causes*

Can humans cause aridity?

Humans have been able to adapt to aridity from the very beginning by accepting the demands it made on them. They did this either by abandoning sites that were perceived as being impossible to live in, thereby creating void or a "desert" in the human sense of the term, or by using the possibilities offered by the scanty vegetation and limited water resources. After a long human journey (Chapter 1 in this book and Volume 3, Chapter 3), a "desert civilization" was forged over centuries based on an intimate knowledge of the environment and its limitations, whose population was small and scattered. On the one hand, the civilization was made up the "lords of the deserts", the nomadic pastoralists who reared camels or nomadic horsemen; on the other hand, there were the sedentary populations. This infrequent, "molecular" occupation of the land, by mobile or fixed populations placed no great strain on resources or on available space. Apart from a few dramatic episodes, which were also transient and located mostly along the borders of the steppes than in the actual deserts (e.g. the large-scale migrations; Attila the Hun, about whom it was said, "no grass grew where he had passed"; brutal conquests by Genghis Khan and Timur Leng, the Turks, the tribal wars in Africa and America) humans who lived in the arid regions very rarely caused any damage to the environment. In this sense, it could be said that desert civilizations were an ecological success!

However, things changed with the onset of the colonial era (16th Century to the early 20th Century) and worsened during the industrial era (19th–20th Centuries). The increase in the number of ways one could directly act on the environment reached a point where modern life seemed slowly to emerge as a threat to the arid world. The intrusion of foreign civilizations equipped with powerful tools, who were more intent on what the land would yield than on preserving it, combined with the rising needs of a growing population and the problems posed by frequent and prolonged droughts, periodically plunged the arid world into crises, sometimes even leading to famine. The socioeconomic factors that played a role in these crises have often been denounced. The population growth placed an excessive strain on grazing, and the expansion of agriculture was not viable given the fragile vegetation, poor soils and precarious water resources, as in the Sahel for example. Resources were badly managed either due to carelessness or ignorance about the geodynamic restrictions of the environment, as was the case in North Africa. Misguided attempts to introduce incompatible agricultural practices, such as the systematic mechanization of processes or extravagant irrigation, provoked disasters such as those around the Aral Sea, for example in Uzbekistan, where there was salinization around the oases and in the water used for irrigation. But we also cannot forget the role played by politics and the rise and fall of States, whether colonial or independent: wars, the establishment of more or less impenetrable borders, sedentarization that was more or less forced, the expansion of export crops and overexploitation of resources. There was also the megalomania of leaders, especially when it came to engineering and water resources, as in Egypt or Libya, for instance, where the vulnerabilities pushed leaders to create utopian projects in the 1970s and 1980s (which happened to be the driest period of the century), which were overambitious for such a vulnerable ecosystem.

Whether taken in isolation or together, all these human interventions could have been the root cause for a series of degradations leading to aridity. Let us first look at vegetation. The gathering of firewood, clearing of land, and excessive and selective grazing by herds of animals all placed an immense strain on the biomass and biodiversity. Denuded and trampled underfoot, destabilized by plowing, the soil, which was already quite thin, was further exposed to wind and water erosion. Fields, villages, sometimes even entire oases, faced the threat of sandization. Rainwater, which would run off and form sheets on the glacis filled wadis and would cause sudden floods. Infiltration decreased as a result, limiting the recharge of groundwater reserves, while evaporation, which was exacerbated by denudation and irrigation, caused salts to rise to the surface. Urbanization, even if scattered, was itself a factor causing aridity through deforestation and overgrazing around urban settlements, as well as through the rise in artificial, impermeable surfaces and architectural obstacles: in 40 years, the city of Nouakchott, in Mauritania, suffered from both sandization as well as flooding.

While aridity from natural causes is one thing, it is quite another to have aridity caused by human activity. While it is not always easy to tell them apart, a lot of the degradation and expansion of arid regions are indisputably caused by humans. In any case, the scale is not the same in terms of space and time: transformations in global climate take place over centuries or even over millennia and across continents. Droughts can be measured in years and over a particular region. When produced by natural causes, climatic aridification is a natural hazard that is produced in a slow and oscillating process; aridification through human causes, however, manifests as a socioeconomic phenomenon and is almost catastrophic.

The arid world suffers (even more than other climatic regions, given their greater fragility) from the negative effects of population and economic growth around the planet. However, in keeping with the land use in these regions, the effects are intermittent or, at least, limited. Further, they have a greater impact on the sub-arid or sub-humid borders, than on the arid or hyper-arid heart of these regions: it is hard to desertify a desert. Isolated pockets or patches of aridity first appear, giving the steppes a speckled appearance. These patches then combine to form bare stretches that eventually become one with the desert. This phenomenon, however, is not always irreversible. The Mongolian steppes, ravaged by the warriors, always re-formed. The bare dunes in the Sahel began to once again sport green cover after the long drought from 1970–1985. Battlefields with devastated land, abandoned after wars, saw the re-emergence of life in Tibet, Ethiopia and Kuwait. There have even been sightings of camels grazing on the sites of atomic bomb tests in the Sahara (In Ekker). The flourishing of protected lands shows that an environment can be rehabilitated. However, if hydraulic projects and fields are no longer maintained following emigration, if sedentarization can no longer compensate for the lack of grazing resources for nomads, and if the pace of devastation is greater than that of restoration, then we may arrive at an irreversible situation, producing true aridification.

Indeed, humans can only impact global climate through industrial pollution and, indirectly, through deforestation and clearing. However, they can also cause significant modifications in geodynamic hydrological parameters that influence climate or depend on climatic factors. Even if humans do not create aridity, they can provoke it, amplify it and cause it to extend well beyond its apparent limits. In the long term, desertic features may appear outside deserts through the overall effects of aridity: uncontrolled processes in terms of excessive variability and increasingly harsh extremes, a reduction in water capacity, in biological productivity and biodiversity, an obstacle to development, producing impoverishment, abandonment and, eventually, a desert.

3.3. Climatic factors and the numerical expression of aridity

Aridity is produced through the combined action of several physical and biophysical factors, modulated by more or less restricted geographical circumstances. What is important is the overall negative balance, showing a deficit of water that is effectively available to the vegetation, depending on whether the water received stays on the soil surface, infiltrates the soil, runs off, or evaporates, and also depending on whether the region is in a hot land or cold land, located on a mountain or a plain, along an ocean or inland in a continent. Aridity is "the characteristic of a climate resulting from precipitation that is insufficient to maintain vegetation" (WMO 1966).

It is a climatic situation with significant biological consequences, a limiting factor that hinders the functioning of ecosystems and human activity. Thus, it is of primordial importance in general climatology as well as agroclimatology.

3.3.1. *Precipitation*

When we speak of a desert climate, the first thought that comes to mind is the absence or rarity of rainfall and that the climate is as it is because of this absence or scarcity. And while in reality the situation is more complex, poor rainfall is indeed a major feature of aridity.

3.3.1.1. *The difficulty of measuring rainfall in the desert*

It is never easy to measure rainfall, nor to interpret the measurements, and this is even harder in deserts than in other places. For a long time, researchers had to be content with observations provided by travelers and measurements in the soil at observation stations that were few and far between, set up at military or administrative posts, in the oases, in agricultural or manufacturing or technical centers, and manned by staff who may have been devoted to the cause but were not permanent nor always very qualified. As long as people only wanted to study the quantity of water received on the soil, in terms of millimeters per unit area, the simple and venerable tool of pluviometry was adequate. Along with the thermometer, pluviometry was often the sole basis of observations made by pioneers in this field. Exposed to every wind and all possible desolation, this study required a lot of care, something volunteers often did not have either the time or the inclination for. And the effect of this can be seen in old results. Picturesque legends on the use of pluviometers and the recording of data traveled through all the deserts of the world. Consequently, these stations were improved, better managed, complemented by the installation of recording devices in sites that were hard to access, and organized into networks, although these often did not always form a homogeneous whole.

In any case, the oldest stations are hardly more than 100 or 130 years old and are often less than 50 years old. Thus, it is hard to find any series of observations going back a long way and even where older data are available there are often sampling gaps or discontinuities in processing the data. Under these circumstances, it is quite difficult to carry out comparisons or to appreciate local nuances in the data. These gaps must be filled, the series must be homogenized, using various mathematical or statistical methods, research must establish the best reference periods, which make it possible to integrate the largest number of posts for the longest common duration, and finally, through extrapolation or through the modeling of contour lines, we must draw isohyet maps, with lines connecting points that are found to receive the same quantity of precipitation.

Researchers today posses more sophisticated means of investigation (Bardinet 1994). The most common of these is remote sensing via satellites, which allow researchers to explore a field extending over several thousand square kilometers using a single image. Remote sensing provides data on air moisture, nebulosity, the trajectory of pluvial zones and of storms. Because of IR technology and RADAR, the satellite can gather valuable information on the state of the plant cover, in all its diversity and seasonal variations, on the albedo, and on the water content and thickness of soil. However, these are indirect measures that offer more information on phenomena prior to or following rainfall, rather than information on the rainfall itself. If we wish to take the analysis further and obtain hard data on the quality and effects of the rain, we must return to measurement stations. In order to bring together all these elements, we must have the appropriate equipment, automated to a large extent, or use circuitous methods such as rain simulation. Operated from traveling, mobile installations, these simulations artificially create rainfall of a given intensity over the different types of soils and slopes present at the site. All of these methods require knowledgeable staff, who are not easy to find in deserts.

As we can see, despite technological advances and remote sensing methods, ground tracking systems remain an essential source of information in documenting pluviometry.

3.3.1.2. *The quantity of rainfall*

Let us look at some of the meteorological stations that helped construct the *Map of the world distribution of arid regions* (Meigs 1977–1979) and let us classify them as given below (Table 3.1). This is done first on the basis of pluviometric patterns, i.e. the distribution of rainfall through the year. The stations are then grouped geographically: Northern Hemisphere and Southern Hemisphere; Africa, the Middle East, Central and Eastern Asia, America and Australia. Finally, within each group, the stations are arranged in increasing order of the total annual precipitations (P).

On the whole, the total annual precipitation, or quantity of rainfall, is quite low. Only five stations received more than 200 mm of rain, and all of these are in the Southern Hemisphere. Twelve stations received less than 100 mm of rain and, consequently, are considered truly desertic stations: four of them are found in the Sahara, including the driest of all stations, In Salah, which measured only 14 mm of rainfall, closely followed by Swakopmund in Namibia (19 mm).

Stations		Latitude	Altitude	P	nH	H	mH	nS	S	mS
			(m)	(mm)		(mm)			(mm)	
Rainfall at peak summer										
Atar	Sahara	20°31′N	225	93	2	61	A-S	10	32	O-Jt
Nouakchott	Sahara	18°7′N	5	141	2	110	A-S	9	20	O-J
Khartoum	Sahara	15°37′N	148	180	2	144	Jt-A	8	8	O-Ma
Jacobabad	Pakistan	28°17′N	56	89	2	46	Jt-A	10	43	S-J
Kucha (Kuqa)	China	43°40′N	1060	79	1	33	J	10	28	A-Ma
Astrakhan	Russia	46°22′N	< 0	122	3	70	Ma-Jy	1	8	Jv
Yinchuan	China	38°28′N	1164	145	4	108	J-S	6	4	O-M
Urga	Mongolia	47°55′N	1449	189	3	166	J-A	8	23	O-Ma
Tuléar	Madagascar	23°20′S	6	348	5	275	N-M	1/3	7/18	Av/Jt-S
Birdsville	Australia	25°54′S	43	161	2	53	J-F	1/2	9/12	Ap/A-S
Alice Springs	Australia	23°49′S	545	246	5	150	N-M	1	6	S
Rainfall at the height of winter										
Palmyre	Syria	34°36′N	700	148	2	67	D-Jv	5	5	J-O
Baghdad	Iraq	33°20′N	60	167	4	125	N-F	6	7	Ma-O
Jask	Iran	25°45′N	10	108	3	76	D-F	8	18	Ap-N
Noukous	Turkestan	41°5′N	112	97	1	27	M	7	25	Ma-N
Southern Cross	Australia	31°13′S	356	279	6	187	M-A	0	-	-
Rainy season with two maxima										
Mary (Merv)	Turkestan	37°7′N	239	128	2/1	52/23	M-Ap/D	5/1	6/2	J-O/D
Las Vegas	Nevada	36°10′N	648	105	0	-	-	4/3	21/21	M-J/S-N
Phoenix	Arizona	33°30′N	334	177	2/1	50/24	A-S/D	3	14	Av-J
Tarcoola	Australia	30°43′S	118	170	1	21	F	0	-	-
Meekatharra	Australia	26°35′S	517	209	3/2	83/60	Jv-M/Ma-J	4	22	S-D
Bourke	Australia	30°5′S	108	346	10	306	O-Jt	0	-	-
Santa Cruz	Argentina	50°1′S	11	193	2/1	42/23	Av-Ma/D	2	15	S-O
Random rainfall										
In Salah	Sahara	27°12′N	280	14	0	-	-	12	14	Jv-D
Tabelbala	Sahara	29°27′N	588	31	0	-	-	12	31	Jv-D
Cairo	Egypt	30°5′N	23	32	0	-	-	12	32	Jv-D
Aden	Yemen	12°45′N	10	45	0	-	-	12	45	Jv-D
Kachgar	China	39°30′N	1418	49	0	-	-	12	49	Jv-D
Swakopmund	Namibia	22°42′S	6	19	0	-	-	12	19	Jy-J
Potrerillos	Chile	26°30′S	2850	44	0	-	-	12	44	Jy-J
Lima	Peru	12°4′S	128	39	0	-	-	12	39	Jy-J

P, total annual precipitation; nH, number of "wet" months (> 20 mm); H, the total precipitation during the "wet" months; mH designating "wet" months; nS the number of "dry" months (< 10 mm); S the total precipitation during the "dry" months; mS designation of "dry" months. Months: Jv, January; F, February; M, March; Ap, April; Ma, May; J, June; Jy, July; A, August; S, September; O, October; N, November; D, December.

Sources: Map of the world distribution of arid regions, MAB 7 technical notes (Meigs 1977–1979), and others.

Table 3.1. *Arid regions – precipitation*

Despite the fact that the rainfall is insufficient, studying the annual quantity of precipitation does offer some relevant information about the global distribution of deserts and enables us to make certain observations: 22 out of the 31 stations are located in the lower latitudes, between 12 and 35° North or South. These are the zones that receive maximum sunlight, corresponding to two chains of tropical anticyclones on either side of the equator. Nebulosity is lower here than anywhere else (< 5/10 on average), the temperature is very high and relative humidity is relatively low (<50%) especially in the Northern Hemisphere. The nine other stations, apart from Santa Cruz (in Argentina) all lie in the interior of continents in the Northern Hemisphere, between 35 and 50°, where the ratio of land cover to ocean cover exceeds 50%. The relative humidity rarely exceeds 60%, but is maintained in winter by low temperatures and in summer by high evaporation. Nebulosity exceeds 5/10, which thus also reduces sunlight and limits the temperatures. The sky is often covered but precipitation is rare and less than 200 mm everywhere, also partly occurring as snow.

These annual data, however, must not be given undue importance. These are average figures, established over a long period, and mask the realities that make up the climate that is experienced in the field and which are even more illusory when aridity is greater. The precipitations are experienced differently by living creatures depending on whether they occur in summer, in winter, as rain or snow, through the year or in specific seasons, or as isolated rainfall between prolonged droughts. To say (as people still often do) that the desert begins only beyond a specific annual pluviometric value (generally 100 or 200 mm) is only a rough approximation that cannot be selected as the only expression of aridity.

3.3.1.3. *Seasonal distribution patterns*

All things said and done, rainfall does occur in deserts. And as drought or dryness is the common lot of all deserts, it is the rainfall that differentiates one desert from another. By pure convention, months in which there is over 20 mm of water are considered "wet" and those that receive less than 10 mm of water are considered "dry". In general, rain is spread out over one or more "wet" seasons, that alternate with "dry" seasons, which are of varying durations. But the fact remains that precipitation, whatever the quantity, is spread out over the year, with poorly marked and uncertain maxima (Table 3.1) or occur irregularly with no predictable dates.

The first group includes stations where the maximum rainfall occurs in summer

This can be further subdivided into two geographically distinct groups:

The subgroups with deserts in tropical regions located between 12 and 35° of latitude North and South. In winter, this area is subject to dry winds from the subtropical anticyclones (trade winds) and in summer they have incursions of humid maritime air (monsoons). There may be significant precipitation (> 100 mm) that occurs between the end of spring and the beginning of autumn, when the Sun is close to the zenith. This is the case in the Mauritano-Sudanese Saharan region (Atar, Nouakchott, Khartoum), southern Africa (the Kalahari), Madagascar (Tuléar), Mexico, Argentinian-Paraguayan Chaco, Central Australia (Alice Springs, Birdsville) and, closer to the equator, in Northeast Brazil. This is also the case in countries where the Indian monsoon blows: the Indus Valley (Jacobabad), the Thar Desert and the western Deccan region. The "wet" period is relatively short (2–5 months) and of varying intensity, depending on local conditions. The "dry" period, sometimes interrupted by a slight secondary maximum (Tuléar, Birdsville), is very long (8–10 months) in the Northern Hemisphere and always shorter (1–4 months) in the Southern Hemisphere.

The subgroup of continental deserts located in the middle latitudes between 35 and 50° extends across Central and Eastern Asia from the Caspian region (Astrakhan) to Mongolia (Urga), and over the Interior West of the United States. Exposed to the cold winds from the Arctic or from the Siberian thermal anticyclone, winters here are dry and harsh, with conditions being further intensified by altitude and continentality. In Asia, the dry season (or the less humid season) is long and frozen (8–10 months in Kusha and Urga). This is also the season in which the maximum precipitation is in the form of snow, especially in March and October. The maximum is situated between May and September, the same time as the high temperatures. The annual quantity is not much (< 200 mm) but it increases as we get closer to the Pacific coast (Urga, Yinchuan). The situation in the inland deserts of North America (Utah, Colorado) is almost equally as harsh. In addition to the cold, dry conditions of winter there is also deficit rainfall in summer, as a result of the coastal mountain chains blocking maritime air and preventing its entry into the region. What precipitation there is comes chiefly from local storms. With the *chinook*, the desert spills over as far as the wide piedmont plains of the Rockies, between Nebraska and New Mexico. The maritime Atlantic air, which comes in from the Gulf of Mexico, dies away before reaching the Rockies, while the cold winds from the north maintain the dry and prolonged winters.

The second group consists of stations where maximum rainfall occurs in winter

Most of them lie in the subtropical latitudes between 30 and 40°, chiefly in the Northern Hemisphere. In Afrasia, these occupy a band of regions from Morocco (and from Spain) to Turkestan (Noukous) across North Africa, West Asia (Palmyra, Baghdad) and Iran (Jask), up to the fringes of what we call the Mediterranean climate. The chief characteristic of this climate is the existence of a hot and dry summer, which is physiologically demanding for vegetation, as it compounds the heat retention that occurs in winter. In summer, as the subtropical high pressure zones move up in latitude, they prevent the free circulation of MPAs which bring in the winter rains. The "wet" season occurs sometime between the end of autumn and the beginning of spring. The "dry" season lasts 5–8 months, between April and

November or even up to December. This season tends to be longer toward the south, where the return of the MPAs is slow, and toward the east, where the activity of pluviogenic systems dies out as they progress over the continent. These internal degradations lead to increasingly desertic situations, further aggravated in Asia by the fragmentation of relief features. Weather systems with droughts in the summer and rains in the winter also exist in the Cape province in South Africa, southwest Australia (Southern Cross) and in the western regions of North America (California, Nevada).

A third group consists of stations where the rain patterns present two clear maxima, which are more or less distinct

This group forms a sort of transition between the two previous groups.

Some of these stations have a primary maximum in the summer, as in tropical lands, but also a secondary maximum toward the end of autumn, as in the subtropical countries. These regions (with the two maxima) can be found between the latitudes 35 and 0°, from Pakistan to the Red Sea and the Horn of Africa, on the eastern edge of the Indian monsoon, which brings in summer rains, and also in southeastern Australia (Bourke) and southwest Australia (Meekatharra), northwest Mexico, lower California and in Arizona (Phoenix). The total precipitation is relatively higher throughout these regions (> 100 mm). The dry periods that intervene occur chiefly in spring and again, to a lesser extent, in autumn. These dry periods are short (2–4 months) and may sometimes be entirely absent: there is not a single month that sees less than 10 mm rainfall in Tarcoola, and less than 20 mm in Bourke.

Other stations, between latitudes 35 and 50°, have a principal maximum in spring or in autumn, as in the Mediterranean countries, or in summer, as is the case in continental countries, with a secondary maximum in winter. The dry seasons are distinct, especially in stations that are clearly continental or sheltered from maritime winds. For example, Mary, Central Asia, where there is a dry season for 5 months between June and October, and another month in December, or North American stations such as Las Vegas in Nevada, where the dry season lasts for 4 months between March and June and then 3 months between September and November. A maritime version of these stations can be found in Santa Cruz, in Argentinian Patagonia, where there are two maxima, in autumn and in summer, with the dry season reduced to 2 months at the beginning of spring and a secondary, weaker season at the start of autumn.

The final group of stations, which are among the most arid, comprises stations where the rainfall pattern is random

There is no clear, established maximum, nor a specific season. Dry conditions prevail for most of the year and can even last several successive years. Local conditions (continentality, rain shadow, cold coastal currents) generally play a large role in these regions. This is the case with multiple sites in Central Sahara (In Salah), septentrional Saharan regions (Tabelbala), the eastern Saharan region (Cairo), in southern Arabia (Aden), and in the uplands of Central Asia (Kashgar). In the Southern Hemisphere, these include the Andes north of Chile (Potrerillos), the Peruvian coast (Lima) and Namibia (Swakopmund).

3.3.1.4. *Concentration and variability*

On a global scale, pluviometry in deserts may thus appear to be an extreme degradation in regional pluviometric systems. However, once again this is a rough and simplistic view of things. There is such great diversity across space and variability across time, that it is perfectly possible to travel through deserts without ever encountering the conditions described by statistics.

Not all rainfall is similar and it has very different impacts on the surface of the soil and plant production, depending on the case. Precipitation of less than 0.1 mm cannot practically be measured – the samples talk of "traces". When there is rainfall of less than 5 mm, the soil barely becomes moist before water evaporates or, even more, runs off. Rain is only "effective" if it can moisten the soil enough to sustain the plants there. It is even better when it can recharge groundwater tables and increase water levels in wells. Rainfall greater than 5 mm is thought to be able to do this, more or less, though its impact is still variable and depends on the terrain and topography and, above all, on the intensity or concentration of the rainfall, i.e. its volume per unit time. In this regard, we generally consider *daily rainfall*, as it is easiest to measure, even though it is possible to measure instantaneous rainfall using recording devices.

The most favorable rains are fine rains that last for several hours. However, this kind of rain is quite rare, almost exceptional, in deserts, and can be found chiefly in regions fed by humid air: sea breezes, monsoons or disturbances in the "polar front". In most cases, especially in hot deserts, the rains are highly localized and separated by long periods of dry weather. The gap between two consecutive bouts of rainfall can range from several months to over a year. Some authors (Emberger 1938) consider that for a "true" desert, this interval must exceed 12 months. Indeed, there have been dry periods lasting over 15 years in the Sahara, and over 20 years in the Namib and the Atacama. In the Tamarugal Pampa, in the Chilean Andes, there have been only nine instances of rainfall between 1819 and 1911, and there were 30 consecutive years (from 1823 to 1852) that were completely dry.

On the other hand, certain periods of rainfall can result in enormous quantities of water, usually between 30–100 mm and more in a period of 24 h. The maximum observed rainfall can even often exceed the annual average of the station. At the In Salah station, over 66 years, there was one day where the recorded rainfall in 24 h was 35 mm, while the annual rainfall average for that station was 13 mm. At Alice Springs, in Australia, one particular month saw 340 mm of rainfall, while the annual average was 246 mm. This precipitation, in the form of storms, or brief and heavy downpours, is formidable as they could prove to be devastating. The concentration (sometimes over 1 mm/min) is so intense that the soil is unable to absorb the water. A large part of the water received thus flows away (and is unable to recharge the ground water tables), provoking flash floods. To drown in the desert might seem rather ironic. At the same time, the water that flows away creates ravines in soft soil, carries away debris from the glacis and from riverbeds and is a threat to human installations. The greater the aridity, the greater the risk. Beyond a certain threshold, rainfall can move from "effective" to "hazardous".

These unusual pluviometric events are further aggravated by the large variability of precipitation over time. No two years are alike and the difference can sometimes be startling (see map of interannual variability; Mainguet 1995, p. 94). At Alice Springs, although the annual average is 246 mm, the actual received rainfall is often below 100 mm; however in 1974, over **1 m** of rain was recorded. The same station that once saw 340 mm of rainfall in a month in 1 year received only 5 mm of rain in the same month the following year. At Swakopmund in Namibia, for the first third of the 20th Century, the annual total varied from 1 to 148 mm (out of which, 112 mm was received in a single month). In Arica, on the northern coast of Chile, one year recorded 100 mm of water received from precipitation and the next 9 years were completely dry. Expected rainfall does not occur, or does not occur when expected, arriving either too early or too late for the farmers in the oases, or for pastoralist communities in the steppes and the desert. Dry years follow wet years, like how the gaunt cows followed the plump cows in the Biblical story, but there is no set pattern or frequency. As a result, in highly arid regions, the average rainfall statistics have no real significance and any long-term prediction is illusory. Indeed, one of the essential features of a desert is the unpredictability of rains.

Although torrential rains can wreak havoc, these are limited in time and space and, on the whole, wet seasons are beneficial for those living in the deserts. The dry periods, on the other hand, can produce terrible hazards and have a much wider range of impact, especially along the borders of the deserts, both in physical and human terms. These are, of course, marked by a lack of water, leading to accelerated and cumulative degradation of the green cover and agro-pastoral production. Over the last century, the ecological destruction and nutritional loss caused by the combined effects of prolonged drought and an excessive human load on the environment has brought entire populations to the brink of famine, caused numerous villages to almost disappear, and led to the widespread death of livestock, beginning with cows and ending with goats (open grazing).

Although the quantity of rainfall is always very small, rain is far from being an unknown entity in deserts. In fact, it is a major climatic factor that is imperfectly captured by statistical values. Its sporadic distribution over space and its extreme variability over time chiefly characterize the most desertic of the deserts. These traits are attenuated closer to the borders, where weather patterns from neighboring regions gradually take over. Overly long dry periods may, however, disrupt this situation and in such cases the desert can encroach dangerously into surrounding territories. During the second half of the 20th Century, along the Sahelian edge of the Sahara, the 100 mm isohyet shifted southwards by several hundred kilometers latitudinally. The borders of a desert are the most fragile and the most exposed to unforeseeable climatic risks.

3.3.2. *Temperatures*

Let us return to the meteorological stations located in "arid" regions that we saw in Table 3.1, and reclassify them in increasing order of latitude in each of the two hemispheres (Table 3.2).

In deserts located in the lower latitudes of the Northern Hemisphere, the incident energy e_i is higher, due to the close verticality of solar rays falling on this region, the days and nights being almost equal in length, and low nebulosity, which ensures maximum light coming in. Between the latitudes 15 and 35° N, which is where the largest deserts are located, almost 60% of sunlight received at the level of the atmosphere reaches the ground, while the albedo rarely exceeds 40%. The available energy e (excluding IR radiation) is, therefore, largely positive, and this generates high temperatures even in winter. Thus, deserts in the lower latitudes are *hot deserts*.

Annual average temperatures throughout these regions are very high (>20°C), except in the mountains. The stations in the Sahara and in hot deserts in Asia have observed record annual averages: 28°C in Atar (Mauritania), almost 30°C in Khartoum (Sudan), greater than 28°C in Central Arabia, and 27°C in Pakistan, close to the Thar desert. The figures are high even along the coast: 29°C in Aden (Yemen), which is located quite far south, it is true, at (12°46′N), 25°C in Nouakchott (Mauritania) on the Atlantic coast (18°7′N), and 27°C in Jask (Iran) on the Gulf of Oman, despite its higher latitudinal position (25°45′N). These averages have a natural tendency to decrease as we move higher up the latitudes, however, the difference is very small and there are many local exceptions, resulting from the altitude or continentality. For example, the average at In Salah, in the central Sahara (27°12′N and an altitude of 280 m), is 27°C , while in Tabelbala, in the northern Sahara (29°27′N and 588 m), it is 24°C. Similarly, the average temperature is 24°C in Baghdad (Iraq) (33°20′N), while it is 21°C in Phoenix, Arizona (33°30′N and 334 m). For average temperatures lower than 20°C, we must head to the mountains: the Hoggar or Tibesti ranges, mountains in Iran or Afghanistan, and ranges in the American West.

Stations		Latitude	Altitude (m)	$\bar{t}$ (°C)	t_{max} (°C)	M	t_{min} (°C)	m	$t_{max} - t_{min}$ (°C)
Northern Hemisphere									
Aden	Yemen	12°46′N	10	29	32	J	25	D-F	7
Khartoum	Soudan	15°37′N	148	29.5	34	Mi-J	24	F	10
Nouakchott	Mauritania	18°7′N	5	25	29	S	21	D-Jv	8
Atar	Mauritania	20°31′N	225	28	34	J-A	20	Jv	14
		23°26′N	*Tropic of Cancer*						
Jask	Iran	25°45′N	10	27	33	Jt	20	Jv-F	13
In Salah	Algeria	27°12′N	280	25	37	Jt	13	D-Jv	24
Jacobabad	Pakistan	28°17′N	56	27	36	Jt	14	Jv	22
Tabelbala	Algeria	29°27′N	588	24.5	36	Jt	13	Jv	23
Cairo	Egypt	30°5′N	23	21	29	Jt	12	Jv	17
Bagdad	Irak	33°20′N	60	24	35	Jt-A	10	Jv	25
Phoenix	Arizona	33°30′N	334	21	32	Jt	10	Jv	22
Palmyre	Syria	34°36′N	700	20	32	A	8	Jv	24
		35°0′N	*Hot deserts/cool deserts*						
Las Vegas	Nevada	36°10′N	648	19	32	Jt	7	Jv	25
Mary (Merv)	Turkestan	37°7′N	239	15.9	30	Jt	0	F	30
Yinchuan	China	38°28′N	1164	8	23	Jt-A	-9	Jv	32
Kachgar	Xinjiang	39°30′N	1418	12	26	Jt	-7	Jv	33
Noukous	Turkestan	41°5′N	112	12.5	28	Jt	-5	Jv	33
Kucha (Kuqa)	Xinjiang	43°40′N	1060	9	24	Jt	-14	Jv	38
Astrakhan	Russia	46°22′N	< 0	10	26	Jt	-7	Jv-F	33
Urga	Mongolia	47°55′N	1449	-2	17	Jt	-23	Jv	40
Southern Hemisphere									
Lima	Peru	12°4′S	128	19	23	Jv	15	Jt-A	8
Swakopmund	Namibia	22°42′S	6	15	17	Jv-M	13	A-S	4
Tuléar	Madagascar	23°20′S	6	24	27	Jv-F	19	Jt	8
		23°26′S	*Tropic of Capricorn*						
Alice Springs	Australia	23°49′S	545	21	29	Jv	12	Jt	17
Birdsville	Australia	25°54′S	43	23	31	Jv-F	13	Jt	18
Potrerillos	Chile	26°30′S	2850	11	14	D-Jv	8	J-A	6
Meekatharra	Australia	26°35′S	517	22	31	Jv	13	Jt	18
Bourke	Australia	30°5′S	108	20	28	Jv	11	Jt	17
Tarcoola	Australia	30°43′S	119	19	27	Jv-F	11	Jt	16
Southern Cross	Australia	31°13′S	356	18	26	Jv-F	10	Jt	16
Santa Cruz	Argentina	50°1′S	11	8	14	Jv-F	1	J	13

$\bar{t}$, average annual temperature; t_{max}, average temperature of the hottest month (M); t_{min}, average temperature of the coldest month (m); $t_{max} - t_{min}$, thermal difference between the hottest and the coldest months; months: Jv, January; F, February; M, March; Mi, May; J, June; Jt, July; A, August; S, September; D, December.

Sources: Map of the world distribution of arid regions, MAB 7 technical notes (Meigs 1977–1979) and others.

Table 3.2. ***Arid regions – temperatures***

The deserts lying in the middle latitudes are at a disadvantage as the Sun's rays strike them obliquely and because of their long nights in winter (inadequately balanced by the long days in summer). Between latitudes 35 and 50°N, which is where the large Eurasian deserts and many American deserts lie, a maximum of 50% of the energy from the Sun reaches the ground, with an albedo greater than 40%, partly amplified by the high reflectance of snow and clouds (when these are present). The energy balance, *e*, rarely exceeds 10% and is even negative during winters. With quite a significant difference in the average temperatures in different seasons, lower than 10°C in winter and higher than 20°C in summer, the deserts in the middle latitudes are cool or temperate (konwn as cold winter deserts). It is only the polar deserts that are truly *cold* deserts.

The annual average temperatures are all lower than 20°C. Indeed, in Las Vegas (Nevada) located at 36°10′ N and at an altitude of 648 m, the average is as low as 19°C, while it is 15.9°C in Mary (Merv) in Turkestan, which is still relatively south (37°7′N) and at a low altitude (less than 300 m). On the other hand, the average temperatures do not exceed 12.5°C in Noukous, located in the lower reaches of the Amu Darya (41°5′N), and only go as high as 10°C in Astrakhan, in the Caspian depression at 46°22′N, lying below sea level. Further east, in addition to the low energy balance there are the factors of altitude and continentality: 12°C in Kashgar (39°30′, 1418 m), to the extreme west of the Tarim basin; 9°C in Kusha (or Kuqa) (43°40′N, 1060 m), on the southern slopes of the Tianshan, 8°C at Yinchuan (38°28′N, 1164 m) on the Yellow River, along the edge of the Alashan desert. Urga, in Mongolia (47°55′N and 1449 m), holds the records for the lowest average temperature at −2°C.

There are no cold winter deserts in the Southern Hemisphere except in the Andes and the extreme south of Argentina: in Potrerillos, in the Chilean Atacama, at an altitude of 2,850 m and at the latitude 26°30′S, the annual average is 11°C; in Santa Cruz, on the Atlantic coast of Patagonia, located at 50°2′S, the annual average is no higher than 8°C. However, no southern continent other than South America stretches so far toward the pole that there is a significant drop in temperature, and the other continents are also less massive than in the north. Only Australia extends far enough into the tropics to have a large desert. Here, while the average annual temperatures are high, they are lower than those of the deserts in the Northern Hemisphere: 21°C at Alice Springs, at the heart of the continent (23°49′S, 545 m), 23°C in Birdsville, to the east (23°54′S, 43 m), 19°C in Tarcoola, in the south (30°43′S; 119 m). On the other hand, the oceans occupy a greater proportion of this hemisphere, with 81% of the hemisphere comprising of oceans. The Antarctic ice sheet also takes up a significant portion of the hemisphere. Both these entities play an important role in regulating the climate. The ocean has a much higher capacity for absorbing radiation than the continents (this is why they generally appear black in satellite images). The albedo decreases correspondingly, barely going higher than 10%. As a result, the energy balance *e* is positive everywhere, and the ocean is an enormous reservoir of heat and energy. This energy is chiefly dissipated through evaporation from the surface of the water, which is the principal source of atmospheric humidity, with greater nebulosity than in deserts in the Northern Hemisphere. This energy also drives thermal transfers and mechanisms whose consequences for the climate are not yet understood. For example, maritime air masses traveling over the coasts and, especially, currents and the displacement of marine water (like the notorious El Niño event) or the upwelling of colder water along the coasts, which are responsible for the harshest coastal deserts, such as Swapokmund, Namibia (15°C on average, located at 22°42′S) or in Lima, Peru (12°C at 12°4′S and an altitude of 128 m), while coastal stations that are sheltered from the maritime winds, like Tuléar, Madagascar (23°20′S) record temperatures that are more commonly observed at those latitudes (24°C). These aspects are the chief causes behind the unique features of the arid zones in the Southern Hemisphere.

Just as rainfall statistics do not offer a complete picture, the annual average temperatures alone cannot define the climatology experienced in deserts. These are somewhat abstract statistical values, which do not adequately represent the nuances or irregularities over time, which are more directly experienced in the field.

3.3.2.1. *Seasonal variations*

The seasonal differences between the average temperatures for the hottest months and the coldest months are always greater than 10°C, with the exception of stations in the lowest altitudes (7°C in Aden, 8°C in Lima), along the coasts (8°C in Nouakchott, 8°C in Tuléar, only 4°C in Swakopmund) and in the high tropical mountains (6°C in Potrerillos), where winters are cold and summers are also cool. In general, along the same latitudes, the summers are hotter and the winters are colder in deserts than in humid regions. These are ordinary climatological features that have been exaggerated by the aridity. Following the apparent movement of the Sun and the actual movement of the thermal equator, the highest temperatures occur between May and September, especially in July–August in the Northern Hemisphere, and between December and February, mainly in January, in the Southern Hemisphere. The lowest temperatures occur between December and February, most often in January, in the Northern Hemisphere, and between June and September, usually in July, in the Southern Hemisphere.

In hot deserts, the winter averages are warm (10–25°C) and the summer averages are very hot (30–40°C; 37 at In Salah in July, 35°C in Baghdad in August, 32°C in Phoenix in July). Consequently, the significant difference between the summer and winter average (25°C in Baghdad, 24°C at In Salah; 22°C in Phoenix) are chiefly due to the summer temperatures. The difference is slightly smaller (less than 20°C) in the Southern Hemisphere, which is more maritime, even far inland (17°C at Alice Springs). The difference is also smaller at high altitudes as summer temperatures are lower in these areas (at Potrerillos, at 2,850 m, the average temperature is 14°C in January, and 8°C in July, with a difference of only 6°C). In cold winter deserts or cold continental deserts, however, there are sharp contrasts, with the winter average being quite low (<10°C or even below (0°C), while the summer average is hot, >15°C and even >25°C). The differences are large everywhere, even at raised altitudes: over 25°C (30°C at Mary, at a height of 239 m; 33°C in Kashgar, at 1,418 m; 40°C at Urga, at 1,449 m). However, in these deserts, the large difference is due to the winter averages (0°C in Mary in February; −7°C in Kashgar, −14°C in Kusha, −23°C in Urga in January), as the summer averages are not much lower than those of the hot deserts (32°C in Las Vegas in July, 30°C in Mary; 26°C in Kashgar; 24°C in Kusha; 17°C in Urga).

However, all these are only average temperatures, often quite far removed from the day-to-day temperatures that are actually experienced by the people in the deserts. The absolute maximum temperatures can sometimes be very hard to tolerate. In summer, in the shade (with no exaggeration) temperatures up to 49°C have been recorded in Australia, up to 56°C in Death Valley in California and up to 65°C in the Sahara (Dubief 1959–1963). Even "cold" deserts in Asia can see summer temperatures between 40 and 50°C. In contrast, winter temperatures often fall below 0°C, even in "hot" deserts. In Sahara, frost is a regular feature of winter, at least in the north, and in the mountains above 1,500 m; and the absolute minimum temperatures can go as low as −7°C in Tamanrasset, at an altitude of 1,400 m, and −10 or even −20°C in the Tibesti, above 2,000 m. In Asian deserts, these minimum temperatures can be even lower: −20 to −30°C in Turkestan, −40 to −50°C in China and Mongolia. Thus, the actual difference in temperatures that humans, animals and plants in the deserts must tolerate can be very large: 50 – 60°C in the Sahara and other similar deserts, and 60–70°C in Asia.

Thermal contrasts, which can often be extreme in space and time (shade or direct sunlight, the varying reflectance of different soil surfaces), can reach values in deserts that are unheard of in regions that are more humid or have greater green cover. In general, these contrasts are more pronounced in winter than in summer, and in cold deserts than in hot deserts. They also reduce considerably along the borders of deserts, which are less arid. Apart from the seasonal variations, there are also notable contrasts from one year to another, depending on the general variations in climate. However, despite everything, these annual differences in temperature are less marked than annual differences in rainfall.

3.3.3. *Evaporation*

A large proportion of water that falls onto deserts returns to the atmosphere through evaporation. This is the phenomenon through which water changes from la liquid state to a vapor state (the comparable phenomenon for water in solid state, such as snow or ice, is called sublimation) and this physical phenomenon is as significant as precipitation. Indeed, in a way, this is the corollary to precipitation, but is more complex and, consequently, even harder to measure and interpret.

This is because, in reality, evaporation consists of two different aspects:

1) evaporation proper, denoted by E, which is physical evaporation from open water surfaces, soil and groundwater;

2) physiological evaporation or transpiration from living things, chiefly plants, as they cover large areas, but also from animals, which are more widely scattered and mobile.

In the 1940s, American climatologist C.W. Thornthwaite (Thornthwaite 1948), proposed the term *evapotranspiration* (ET) to denote the overall loss of water from the earth's surface into the atmosphere. The cause for evaporation is, once again, solar energy, through the action of air temperature along humid surfaces and around living organisms.

Physical evaporation is a direct result of temperature and increases as temperature increases. It is at a maximum during the day, in hot regions and highly insulated regions. However, evaporation also depends on other factors that are present and active to differing degrees in deserts.

The first of these is humidity in the air, which governs how much water vapor can be absorbed. When relative humidity is low, the atmosphere can absorb large quantities of water vapor. This capacity decreases as the air gets saturated. We have already seen that air in the deserts is far from being saturated, which makes the atmosphere particularly drying. Another contributory factor is wind, which is a displacement of air and a near-permanent feature in the desert. When there is a rising air current, on the windward side of a relief feature, for example, air becomes less dense and cools down, which increases its relative humidity, reduces or stops evaporation, and may even lead to precipitation. When air descends, however, it becomes compressed and grows warmer, which reduces the relative humidity and increases its drying power (the Foehn effect). A strong wind can agitate, break up and pulverize liquid surfaces; at the same time, it blows away humid air and causes the dry air that previously lay above this layer to prevail. The greater the wind speed, the more intense this agitation and quicker the replacement of humid air by dry air, which is conducive to, and maintains, evaporation. Finally, plant life also plays a role in evaporation, although in a slightly counterintuitive manner. On the one hand, every plant, bush or shrub creates a protective microhabitat around itself, which tends to conserve a little humidity and coolness. On the other hand, given its expanse, vegetation contributes more to evaporative processes than other living organisms through its own physiological activity.

Another factor to consider, of course, is that there must be water for evaporation to occur. Deserts have few evaporating surfaces and these are usually ephemeral, or artificial and highly localized: occasional or temporary wadis, scattered bodies of stagnant water (puddles, dayas, sebkhas) that are quick to disappear, and exceptional lakes, dams (but only on large rivers), irrigation canals and irrigated soils. Water from the soil does also evaporate, but the effect of this is varied depending on the porosity of the material and, in any case, this phenomenon grows weaker deeper in the soil: its effects can no longer be observed beyond a depth of 4–5 m. Evaporation is most intense when the Sun and wind follow a short but heavy shower.

The contribution made by transpiration to the total evaporation is, *a priori*, smaller in deserts than in other areas, due to the scattered nature of the vegetation, with the exception of oases. Nonetheless, it is neither negligible nor easy to gauge. The circulation of water through living things involves a number of variables that are difficult to measure. Thus, in most cases, researchers must be content with a more or less approximate estimation of overall evapotranspiration. Thornthwaite identified two kinds of characteristic quantities for doing this:

1) *potential evapotranspiration* (*PET*), which is the maximum amount of water that would evaporate from an unbroken surface of water, or soil completely covered by continuous vegetation, with the supply of water in both cases being unlimited. This is a theoretical concept, independent of the nature of the soil and vegetation and where the only limiting factor is solar energy, i.e. temperature. In general, this concept is sufficient for defining small-scale climatic units;

2) *actual evapotranspiration* (*ETa*) is the actual evaporation that is measured as having occurred over a given time, from a given surface, taking into account the soil, water resources, green cover and the ability of the plants to regulate loss through transpiration. It is used in local agroclimatology operations and to monitor large-scale farming.

Several evaluation methods have been developed to measure the extent of evaporation. These differ in usability and complexity. Some are based on direct observations, using instruments called evaporimeters, which are simple, but whose results may not be fully accurate. These can be open pans containing water (e.g. the Colorado Sunken Pan), tubes with porous connections (such as the Piche evaporimeter) or Lysimeters (to measure evaporation from soil). All of these, however, only measure physical evaporation. Those that measure evapotranspiration use formulae to establish relationships between the climatological measurements made on the ground (such as solar energy, temperature, humidity and wind speed (Penman 1956), overall radiation, temperature and relative humidity (Turc 1961)) or empirical measurements from laboratory or greenhouse experiments: air temperature, humidity, precipitation, water retention by plants and in the soil (Thornthwaite 1948).

The results that are obtained are presented statistically per day, per month, or for a number of years. However, they do contain some flaws. First of all, the values measured in this manner are the average values of potential evapotranspiration (PET) which, given the scarcity of water, is always greater than actual evapotranspiration (ETa) in the deserts. Further, these methods are not all equally reliable in all contexts. Turc's formula, for example, is better suited to humid regions and for agronomic purposes, while Thornthwaite's formula is more appropriate for dry regions and to distinguish between climates.

In any case, it is very hard to collect usable data from deserts. Today, it is possible to remedy the situation to some extent using IR remote sensing, which gives us an estimation (a relative estimation, at least) of temperature, soil humidity and the condition of the vegetation. However, the fact remains that this technique is dependent on access to highly evolved technology, which is still not universally available, and is also based on numerical relationships whose validity has not yet been established and can be improved. However, the technique does make it possible to map the vegetation and the water balance, which can be used to monitor and manage land and water resources.

Evapotranspiration is part of a long chain of physical and biological events (solar radiation, heat, dispersion of water vapor in the air, nebulosity, precipitation, vaporization, transpiration, etc.) that take place at the interface between the atmosphere and the soil and which have unique features in deserts, given the discontinuous green cover. Evaporation is a highly energy-intensive process (it takes about 680 cal to vaporize 1 g of water) and contributes to thermal regulation in an environment by tangibly reducing the temperature over the area in which it occurs. However, in doing so, it removes significant quantities of water (essential for maintaining life) from this environment. This is why the study of evaporation is so important in deserts. Expressed in millimeters per unit area (and despite a margin of error that could be as high as 10–20%) potential evapotranspiration can be compared with precipitation, and may be either in excess of the quantity of precipitation received or lower than it.

Because of the Sun's strength and high summer temperatures as well as high daytime temperatures, and because of the drying atmosphere, the rate of evapotranspiration in deserts is always greater than the quantity of precipitation received, with the exception of certain days or months, whose frequency varies depending on the stations. It is always very high in all hot deserts, with the average annual values being 1,500 mm in Central Australia, 1,700 mm in the Kalahari, 2,300 mm in Arizona and 2,500–3,000 mm in the Sahel. In the Sahara, evapotranspiration values can be extremely high, even exceeding 6,000 mm, and are the highest in the world. This phenomenon is also more intense in inland regions than along the coasts: 6,000 mm in the Libyan desert and in Tanezrouft, 5,200 mm at In

Salah and 4,400 in Atar, as opposed to 2,800 mm at Nouakchott, on the Atlantic coast. The same observation is made in Namibia (1,800 mm at Swakopmund as opposed to about 2,000 mm at Windhoeck), Madagascar (1,300 mm at Tuléar as opposed to 1,600 mm on the plateaus), California (ranging between 1,500 and 3,000 mm). In cold deserts, these values are smaller, on the whole, being less than 2,000 mm. However, these values increase with an increase in average temperature: 1,160 mm at Noukous, Uzbekistan (12.5°C) 1,600 mm at Mary, Turkmenistan (15.9°C) 1,800 mm at Las Vegas, Nevada (19°C). As an essential factor in the water balance (Tardy 1986), evapotranspiration is a key element in the concept of aridity.

3.3.4. *Aridity indices*

The genericness of the term "desert" masks the tangible differences that exist between various regions that are all similarly called "desertic". As we have seen, there is no real "desert climate", instead there are *desert climates*, which share common traits but also subtle differences, which must be qualified and, if possible, quantified with maximum objectivity. It was therefore important to be able to compare these territories that were similar, but had different pluviometric and thermometric conditions. This requirement led to the rise of a variety of numerical expressions that sought to define aridity through a simple ratio between water brought in through precipitation and water lost through evaporation. These mathematical expressions are generally known as *aridity indices* (Martonne 1926).

They involve evaluating water deficit through an equation of the type:

$$P - E \quad \text{or} \quad P/E,$$

where P designates precipitation and E designates evaporation; the larger the deficit, the greater the aridity. However, when presented in this form, the value of these indices will increase in direct proportion to P and in inverse proportion to E. Thus, they function more as indices of humidity or rainfall efficiency, whose values increase as aridity decreases. The difficulty in establishing these indices is chiefly due to the lack of any incontestable and universally accepted methods to measure evaporation, and from the small size of observations networks in the very areas where they would be most useful. The strong links between solar energy, temperature, air humidity, precipitation, water reserves in soil, plant cover and evaporation have all created different orientations for research into indices for aridity.

3.3.4.1. *Indices that consider precipitation and temperature*

Given the overall paucity of data (and of reliable data) on evaporation, the first group of indices use data on temperature instead, which is easier to measure and more widely available. The principle behind these indices is that there is a direct relationship between the temperature and the water lost through evaporation.

– The oldest formula of this kind was proposed by the French geographer, Emmanuel de Martonne. His aridity index was expressed as:

$$I = \frac{P}{t + 10}, \qquad [3.3]$$

where P is the total annual precipitation received (mm), and t is the average annual temperature (°C) (Martonne 1926). The number 10 is added to t in the denominator simply to avoid negative numbers. The formula may be adapted to obtain an aridity index for a particular month, for example, for the driest month of the year by writing:

$$i = 12\frac{P_m}{t_m + 10}, \qquad [3.4]$$

where P_m and t_m are the averages for the month in question, and the whole equation is multiplied by 12 in order to remain comparable with the annual index. Indeed, by comparing the index for the driest month with the annual index, we obtain the following formula:

$$I = \frac{1}{2}(\frac{P}{t + 10} + 12\frac{P_m}{t_m + 10}), \qquad [3.5]$$

which better expresses the limitations and nuances of aridity.

This index was widely adopted by geographers as it was simple and had been calculated and mapped for the entire planet (Martonne 1942). It still drew criticism, however, since evaporation is not simply a linear function of temperature, but also depends on the duration of the dry season, as well as the difference between the highest and lowest temperatures in the area (which can differ greatly from one region to another). The index also cannot be applied in cases where the monthly or even annual precipitation is zero, i.e. P_m or $P = 0$, as is often the case in deserts.

– Thus, in order to partly resolve these problems, the botanist L. Emberger proposed a rain-thermal quotient, expressed by:

$$Q = \frac{P}{2[\frac{M+m}{2} \cdot (M - m)]} \cdot 100, \qquad [3.6]$$

where P is the total annual precipitation, M is the average maximum temperature for the hottest month and m is the average minimum temperature for the coldest month (Emberger 1930); 2 and 100 are coefficients that are added to avoid obtaining decimals and to simplify the quotients. Temperature, here, is represented by the average of both extremes (hottest and coldest) and evaporation is measured by the range of these temperatures. The higher these two values, the greater the aridity in the region.

This formula may be criticized for implicitly establishing a sort of proportionality that has not been demonstrated between the amplitude of the extreme temperatures and evaporation. It was also chiefly designed for the Mediterranean region, where rainfall occurs in winter. Finally, it is also criticized for the same reasons as the Martonne index, being practically inapplicable in the case of a desert climate with low and scattered rainfall.

3.3.4.2. *Formulae that consider precipitation and evaporation.*

A second group of formulae try to use the measures of evaporation itself, by introducing raw values that were directly observed, or values that have been calculated.

– The climatologist Dubief (1950) proposed an *aridity coefficient*, for the Sahara:

$$D = \frac{P}{E_j}, \qquad [3.7]$$

where P is the annual quantity of precipitation and E_j is the daily average evaporation measured by a Piche evaporimeter, which has low reliability. D thus expresses the time (in days) required for the rain that has fallen to be evaporated. In other words, this expresses the duration of the drought at that station.

Indeed, the Dubief coefficient is only an approximation, which does not take into account runoff, infiltration or the physiological evaporation from living things. For these reasons, despite the author's own attempts to substitute P with a more precise measure P_u (for "useful precipitation", i.e. subtracting from the total precipitation the proportion lost through infiltration and runoff), the coefficient has not been widely employed.

– For greater efficacy, the geographer Capot-Rey (1951), while still using the Piche evaporimeter, also tried to take into consideration the conditions in which rainfall occurred and the seasons in which it occurred. In his aridity index:

$$I = \frac{1}{2}(100 \cdot \frac{P}{E} + 12 \cdot \frac{p}{e}), \qquad [3.8]$$

where P is the quantity of annual precipitation, p is the amount of precipitation received in the wettest month, E is the annual evaporation and e is the evaporation that occurred in the wettest month. The multiplier, 100, is only present in order to prevent fractional numbers from being obtained, and the multiplier 12 is present so as to be able to extend the result for that month over the course of the year. We thus compare the annual effective rainfall P/E with the rainfall that occurred during the wettest month p/e, and this ratio grows smaller when this month is hotter.

This index offers the advantage of being able to distinguish between nuances within an arid zone. While it has only been used for local measurements, it did help in drawing up an interesting aridity map for the Western Sahara (Capot-Rey 1951).

– In a similar vein, the Russian climatologist Budyko (1958) defined a *dryness ratio*:

$$D = \frac{R}{L \cdot P}, \qquad [3.9]$$

where temperature, which is a factor in evaporation, is replaced by R, the term for overall radiation ([solar radiation + atmospheric radiation toward the earth] (1-albedo) – IR radiation from the soil), and where P is the annual precipitation and L the latent heat (enthalpy) of vaporization of water. The ratio gives the duration required for solar radiation to evaporate the entirety of water received through precipitation.

The ratio increases as dryness increases, and the map drawn based on these results is able to adequately trace the boundaries between wet regions and dry regions, as well as the climatic contrasts within the arid world. However, given how rarely radiation is measured, it has, so far, only been employed in very limited contexts.

3.3.4.3. *Formulae that consider bioclimatic data*

The arid world is not only made up of bare soil. There are living things that inhabit these regions and the plants, in particular, although they are scattered, play a role in the water balance through their vital functions. Indeed, it can even be said that the green cover is the best indicator of the complex soil climate in a region.

In the early 1930s, the German climatologist W. Köppen pointed out the relationship between spatial climatic areas and large biomes and, therefore, the possibility of defining climatic regions based on the nature of vegetation (Köppen 1931). Thus, along the arid fringes of the Mediterranean, he marked the boundary between the steppe and the desert using a relationship between the sum of precipitation, P, and the average annual temperature, t:

$$P = \begin{cases} 10 \cdot t & \text{if the rains occur in winter,} \\ 10 \cdot t + 70 & \text{if the rains are evenly distributed through the year,} \\ 10 \cdot t + 140 & \text{if the rains occur in summer,} \end{cases}$$

that is, the rains are less effective when the temperature is higher (Köppen and Geiger 1930–1939; Köppen 1931).

In the meantime, the American climatologist Thornthwaite (1948) proposed a double index, which is apparently more rigorous but is hard to establish, while also criticized on certain aspects. The formula used remains similar to the P/E type, but here P is replaced by the difference $S - D$, between the surplus water, S, observed during the wet months, and the deficit, D, observed during the dry months; E is given by the measure of potential evapotranspiration PET.

The difficulty here lies in computing these parameters using the conventional meteorological observations, P and t, and the empirical formulae tested in many agronomic stations in the United States. It is accepted that the potential evapotranspiration represents the water requirement, i.e. the quantity of water required by vegetation to compensate for the loss of water through transpiration. If, for a given month, the water balance $p - etp$ (precipitation – evapotranspiration) is positive, then once the groundwater is recharged, there is a surplus of water, s, which is lost as runoff, and the station is wet. If the balance is negative, there is not enough water to sustain the vegetation, water reserves are exhausted, a deficit d is observed and the station is dry. In order to gauge s or d, therefore, we must take into account the quantity of water that the soil can hold in order to counterbalance the pet. Further, this reserve depends on the nature of the soil and on the type of vegetation. After multiple experiments, Thornthwaite estimated that this water reserve was 10 cm/m of depth in permeable soil, and 30 cm/m in clayey regions. Whether or not the reserves were exhausted in the dry months depended on the density of the green cover and the intensity and duration of the dry period. On the other hand, these reserves were built up again during the wet spells.

The quantities s, d and pet are calculated on a monthly basis. The annual values S, D and PET are obtained by summing up the monthly values, and the annual index can then be derived as follows:

$$I = \frac{S - D}{PET} \cdot 100. \qquad [3.10]$$

The number 100 is used here to eliminate decimal values. The index I is positive (*humidity index, moisture index*, I_h) if the balance is wet, while it is negative (*aridity index*, I_a) if the balance is dry. While complicated – and sometimes disputed – this method has still been widely used internationally. In simplified form, it can be written as:

$$I = \frac{P}{PET}, \qquad [3.11]$$

and it has notably been used to draw up the reference maps for arid regions published by P. Meigs for UNESCO in 1977–1979.

Revisiting Köppen's idea of bioclimatic correspondence, though in a different form and with greater precision, in 1952 the French botanists H. Gaussen and F. Bagnouls developed an original means of expressing the aridity of a station or a region, based on the magnitude, duration and conditions of the dry season (or dry seasons) during the year.

The authors started with a conventional definition of the dry month, according to which "a dry month is that in which the total monthly precipitation, in mm, is less than or equal to double of the monthly average of temperatures in degrees centigrades" (Gaussen and Bagnouls 1952):

$$p \leq 2t. \qquad [3.12]$$

This expression offers the advantage of being able to be represented graphically using an *ombrothermic diagram*, in which each gradient on the precipitation scale (on the right) corresponds to double the value on the temperature scale (on the left). Dryness occurs when the temperature curve is higher than the precipitation curve.

Each of the dry months that are thus defined contain a certain number, n, of days with no measurable rainfall (<0.1 mm), which are modified by various corrective measures, intended to take into account the humidity in the air and occult precipitation. If relative humidity is lower than 40%, the day is considered as a dry day; if it is between 40–60%, it is considered as 9/10ths of a dry day; if it is between 60 and 80%, it is considered 8/19ths of a dry day; if it is between 80 and 100%, it is considered 7/10ths of a dry day and, finally, at 100% relative humidity, the day is considered wet. Days with mist and dew but no rain are considered as only half a dry day. The annual sum of the "biologically dry days" makes up the *xerothermic index*, x, which indicates the total duration of dryness, which can then be distributed over one or more periods over a year using an ombrothermic diagram (Gaussen and Bagnouls 1952).

The calculation of the xerothermic index is a little complicated and we often do not have the data required to compute it. On the other hand, the ombrothermic diagram has been very successful, as the simplest and most convenient means of visualizing the conditions of dryness at a station, regardless of the climate.

The large number of indices available for use demonstrates the complexity and highly intertwined nature of the climatic factors of aridity. Developments in the field of computer science have greatly contributed to making these indices easy to use. However, none of these indices is completely satisfactory, nor can any of them be universally applied. Most of them were constructed on the basis of highly localized regional conditions (the Sahara, the Mediterranean region, the western United States, etc.) and well-equipped stations. The results are debatable, as precipitation, temperature and evaporation are not the only elements of aridity, and because the data are not always comparable. However, there are promising developments in indirect methods related to radiometric measurements via satellites, such as the measurement of albedo and soil humidity. Measurement via satellite is facilitated in the deserts by the extension of bare soil and the widely scattered (or wholly absent) vegetation, which makes it possible to continuously visualize the spatial and temporal variations in the state of the surface and, from this, deduce the state of aridity.

3.4. Nuances in aridity

A climate must be considered to be arid when rains are not adequate enough to supply the water required to continuously sustain vegetation and agriculture without irrigation. Most observers agree that aridity in the desert starts when the steppe gives way to a loose formation which is excessively open, xerophile and almost entirely without trees. Where it is possible to compute them, in the Sahara for example, aridity indices show that this passage from steppe to desert occurs around the values given in Table 3.3. These data excludes the steppe regions from the *desert*. The former are described as *semi-arid* and *sub-arid*, and are marked by alternating dry and wet seasons, by a pattern of annual floods, and by green cover that varies in density but is omnipresent. Arid regions are subject to particularly severe geodynamic conditions characterized by a near-permanent water deficit, irregular and often violent precipitation and runoff, generally a large role played by the wind, as well as the scanty and scattered nature of vegetation and poor soil. However, for all this, the arid world is not homogeneous and is marked by important nuances, which have created conditions that humans have been able to make use of intelligently. Arid regions can thus be subdivided into distinct bioclimatic units, separated by more or less clear boundaries and qualified by differing degrees of aridity.

Author	Formula
Köppen (1931)	$P \leq 10 \cdot T$
de Martonne (1942)	$\frac{P}{t+10} \leq 5$
Capot-Rey (1951)	$I = \frac{1}{2}(100 \cdot \frac{P}{E} + 12 \cdot \frac{p}{e}) \leq 4,5$
Gaussen (1952)	$x \geq 250$
Meigs-UNESCO (1979)	$P/ETP \leq 0,20$

Table 3.3. *The limiting values for aridity indices for the transition between the steppe and a xerophilic formation*

Identifying these units and determining their borders is a challenge. Using only total precipitation or aridity indices alone cannot suffice. The observation network in deserts is too loose and the data it yields too unequal to be exclusively used for establishing an accurate typology and cartography of these sites. Thus, the solution lies more in how we define "landscapes", environments (ecosystems) that are characterized by specific combinations of several criteria that can be observed in the field. These ecosystems are then qualified, wherever possible, using the aridity indices. While being more naturalistic or geographical than physical, this method does yield an acceptable and vivid picture of reality.

The chief criteria involved are as follows.

3.4.1. *Phytogeographic criteria*

The green cover. The state of the vegetation in a region is often presented as being the best indicator of climate and can be immediately perceived by any traveler in the region. There are species, if not genera or entire families, that are unique to arid regions. According to Capot-Rey (1953), the southern border of the Sahara is marked by the appearance of Bur grass (*Cenchrus biflorus*), which is a Sahelian plant, whereas in the north, the septentrional border is marked by date palms (*Phoenix dactylifera*), although this is debatable as the presence of these plants is also moderated by human activities. Gaussen (1957) pointed out that certain species coincide with a specific climatic area. However, the distribution of these species is also governed by a variety of causes: microclimatic, edaphic, physiological, phytosociological and patterns inherited from the past or modified in the present (relics or pioneers), and even human intervention in the form of grazing lands or agriculture. However, the factor that best explains the plant–climate relationship is the grouping of species, as proposed by Köppen (1931). Plant formation through its physiognomy or facies is a feature of the landscape and formations can be visually distinguished from one another. Through its composition, this association of vegetation demonstrates its adaptation to the region's physical, biological and anthropic constraints.

3.4.2. *Hydrological criteria*

These are based on how water flows across the surfaces and in the talwegs. The border between humid regions and arid regions coincides noticeably with the border between regions that have regular drainage (continuously or intermittent) and regions where water flow is occasional or absent. Most deserts also have endorheic drainage toward closed depressions with no contact with the sea, or may have no drainage at all (arheic regions). While it is true that deficiencies of this kind may also exist for various reasons outside arid regions, what matters here is the generic or widespread nature of these conditions in arid zones.

3.4.3. *Geomorphological criteria*

These reveal geodynamic systems whose prevalent features are that there is low chemical weathering, diffused streaming and wind. However, the landforms could be long-lasting, especially when erosion is slow and weak. Thus, the history of a region must be examined: conditions in the past may have created non-arid forms in regions that are desertic today or, on the contrary, events may have caused arid conditions historically to extend well beyond present-day borders of deserts.

3.4.4. *Soil use criteria*

The soil use criteria consider specifically human activities. Many authors (Dresch 1982) consider that arid regions begin where insufficient rainfall limits agricultural production. Consequently, it is imperative that this lack of water be compensated for by a more or less generalized implementation of irrigation measures in deserts. However, this technique implies the existence of usable water resources and is closely linked to the needs and resources of the populations there, i.e. it is linked ultimately to the social conditions. Furthermore, depending on circumstances and time, dry farming can be followed in regions that do not meet the criteria for a desert, while irrigated agriculture can be practiced in regions that are frankly desertic. Given that arable farming is restricted only to oases, the rest of the desert is used for extensive cattle breeding.

Based on these observations, the experience gleaned from the Sahara led scientists to differentiate between various types of desert landscapes (Raynal 1972a; Raynal 1972b), classified in increasing order of aridity, as illustrated by the values of various indices (here, the indices proposed by Capot-Rey (1951), Gaussen and Bagnouls (1952) and UNESCO (1977–1979)).

3.4.5. *Sub-desert aridity*

A region that is classified as *sub-desert arid* is frankly arid, but is not yet completely a desert. This is a region in transition toward becoming a desert, having been a sub-arid region with slight humidity and still considered part of the steppes. Precipitation in these regions is low (<200 mm), but still follows a perceptible seasonal pattern. The green cover is that of a low steppe, woody or herbaceous, but sparse and with stretches of bare soil predominating over the area with plant cover. The species of trees present here (*Pistacia*, *Juniperus*, *Acacia*), which have been decimated by humans for construction purposes and firewood, survive in residual populations in depressions and talwegs, along cones of dejection and open surfaces. Runoff (which is diffuse over the slopes, or concentrated in branched channels, largely the legacy of a less arid past) is infrequent but not unusual. Floods are either allogenic floods or localized floods following showers and are sudden, brief and limited in space. The morphogenetic action of the wind is important, but secondary to the morphogenetic role played by runoff. The wind's actions result in elementary and highly localized dunes. Apart from extensive, semi-nomadic breeding of small cattle, rural activity is similar to that found in oases irrigated by floods and wells.

A region of this kind usually lies between 4.5 and 3 on the Capot-Rey aridity index, 250 and 330 on the Gaussen index (I_G), and 0.20 and 0.12 on the UNESCO index (Table 3.4).

Station	Capot-Rey index	Gaussen index
Laghouat	3.89	290
Er Rachidia	3.30	330

Table 3.4. *Values for aridity indices in sub-desert arid regions*

This is the case of the southern piedmont of the Maghreb-Saharan Atlas Mountains and of the Anti-Atlas ranges as far as the mouth of the river Draa, also called the Pre-Sahara (Capot-Rey 1952; Joly 1953; Joly 1994): (Gafsa, $I_G = 298$; Laghouat, 290; Boudenib, 309; Er Rachidia, 330). More generally, though varied by features unique to each locale, sub-desert arid conditions are usually found along the fringes of the large, flat, tropical deserts (Tripolitania; Sahel, Agadez $I_G = 295$; the Kalahari), along the edges of the Great Basin in eastern Australia, along the intramountainous plateaus in the western United States (Utah, Colorado)), the Andes (Altiplano) and along the foothills of the highest ranges: the Andes (Chaco), the Rockies (the High Plains), Tian Shan and Kunlun (around the Tarim basin).

3.4.6. *Desert aridity*

In regions with *desert aridity*, the rains are even less (in general, less than 150 or 100 mm) and, above all, instances of rainfall are widely spaced out and more irregular. The wind rivals water runoff, or can even eclipse it in the morphogenesis of relief features and the displacement of superficial detritus. The flora and fauna, which is increasingly widely scattered and largely xerophilic, decrease in population as well as diversity. Outside of the few oases equipped with exceptional water resources, human populations only travel in search of unpredictable grazing land or to make traditional, established journeys or one-off journeys. The Capot-Rey aridity index value lies between 3 and 0.30, the Gaussen index value lies between 330 and 360, and the UNESCO index value lies between 0.12 and 0.03. In the largest deserts, it is possible to further identify two significant subdivisions:

– Regions with *moderate desert aridity* or *meso-arid* regions. When compared to sub-desert arid regions, these regions have more pronounced characteristics of aridity, with less humidity. Rains, while irregular, still statistically equal over a fortnight of measurable rainfall every year (>0.1 mm) or three or four days of rainfall greater than 5 mm (Raynal 1972a; Raynal 1972b). The runoff that follows deepens ravines formed along steep slopes on the mountains and erodes softer land in foothills to create *badlands*. The runoff streams feed transient pools of water, spread into shallow and anastamosed channels (*rill wash*) or into sheets along the glacis (*sheet wash*), and it floods enclosed depressions (*dayas*, *playas*), some of which are used for standalone crops (these basins are called *graras maader*). In allogenic water courses, floods still occur upstream, but are very weak, while localized floods are quickly absorbed by the alluvium itself, which is abandoned and reshaped in the talwegs. The wind, which carries sand from the open lands and from the bottom of dry wadis, deposits its load onto large, open surfaces to form simple, isolated sand dunes (barchans and nebkas), or dunes grouped into small ergs. The xerophilic vegetation, which is widely scattered across the rocky interfluves (*Haloxylon*, *Anabasis*, *Cactus* in America) or sandy interfluves (*Aristida*, *Panicum*) shrinks together, on the other hand, in sandy talwegs (*Acacia*, *Tamarix*, *Retama*, *Ephedra*, *Zilla*) and in the shallow clayey-limestone (*Ziziphus, Rhus, Maerua*), where the plants benefit from underflow and groundwater aquifers. There are also short-lived plants that appear after the rains, called *acheb*. The typical values on the aridity indices are: 3–1.25 on the Capot-Rey index, 330–350 on the Gaussen index and 0.12–0.09 on the UNESCO index (Table 3.5). This is what we find on the northern face of the large North Saharan hamadas, from the Tunisian chotts as far as Tafilalt (Rissani, $I_G = 347$) and Seguiet el Hamra, in the Ougarta range, as well as between the altitude of 700 and 1,600 m in the Hoggar mountains, the Tibesti, and the Adrar des Iforhas and Aïr mountains. This class of aridity can also include the Mediterranean edge of the Egyptian Sahara, the Sinaï, the deserts of Israel, Jordan and Syria, Yemen, the septentrional border of the Aral-Caspian region of Turkestan as far as Lake Balkhash, certain regions located at high altitudes in Upper Asia (such as Dzungaria, Tibet and Alashan), inland regions of the Namib, American deserts like Arizona, New Mexico and Nevada, and a large part of the Australian deserts.

Station	Capot-Rey index	Gaussen index
Touggourt	1.95	332
Béchar	1.69	335

Table 3.5. *Values for the aridity indices for a meso-arid region*

– *Accentuated desert aridity* or *plio-arid* or *frankly arid* regions. These regions best represent the desert as imagined by laypersons. The land of wind, large, ancient ochre-colored ergs, surrounded by newer dune formations of a lighter color. These regions also consist of deserts made up of stone, limestone and volcanic regions, with hamadas and intramountainous basins, as well as deserts made up of vast arenas of granitic piedmont. Rainfall (< 50 mm) is reduced to two or three annual showers in which a certain seasonal rhythm can still be discerned, however, over a year can pass without any rains occurring. In these conditions, runoff streams are highly reduced, rare and unorganized or disorganized. These streams are lost in endorheic, tectonic or hydro-eolian depressions, occupied by stretches of silty-clayey floodplains (*playas*) or by expanses of freshwater (*dayas*) or saltwater (sebkhas, chotts, kavirs). The vegetation, which is halophilic, in these regions, takes the place of grazing land for camels and goats. It consists of a few, ancient trees (*Acacia, Balanites*), but above all, herbaceous (*Aristida*) or bushy species (*Calligonum, Traganum, Calotropis*) that grow on sandy ergs, low dunes (*Cornulaca monacantha*, known in Arabic as *had*, much sought after) and sand-covered boulders (*Retama, Genista, Ephedra*). They are also concentrated in talwegs and depressions with more moisture (*Tamarix, Rhus, Suaeda, Salsola*). In addition to this is the *acheb* that grows after the rains.

The typical values on the aridity indices are as follows: 1.25–0.30 on the Capot-Rey index, 350–360 on the Gaussen index and 0.03–0.09 on the UNESCO index (Table 3.6).

Station	Capot-Rey index	Gaussen index
El Golea	1.04	350
Beni Abbès	0.48	352
Adrar	0.30	357

Table 3.6. *Values for the aridity indices for a frankly arid desert region*

Examples of these are the large Saharan ergs: the Great Eastern Erg, the Great Western Erg, the Erg Chech (southwest), the erg Iguidi in the north, the Ouaran ergs, the Azaouad ergs and the Ténéré ergs to the south. There are also the Tademaït, the Guir, Daoura, Draa and Rio de Oro hamadas, and the foothills of the Hoggar and Aïr ranges. Similar conditions are found in western Arabia (Nedjed), Iran and Afghanistan, in the Thar desert, the deserts of Turkestan (Karakum and Kyzylkum), the Gobi desert, the American West (western Nevada, the Mojave, Sonora and Chihuahua deserts, the Bolivian Andes and the Andes in Northern Chile), and Australia (the Simpson, Gibson and Sandy deserts).

3.4.7. *Hyperaridity*

In a *hyper-arid* desert region, "rainfall does not take place every year and may occur in any season" (Emberger 1938). The poor rainfall (< 50 mm) and the lack of a rainfall pattern hamper most hydrochemical and hydrodynamic processes. The wind is the dominant (if not exclusive) morphogenetic agent here (see Volume 3, Chapter 1). Despite this, however, the landforms created by the wind are often very limited in their development as there is insufficient moveable material available locally, due to the slow rate of disintegration of the rocks. The wind's action is largely limited to corrasion or polishing the rocky outcrops, and dusting sand over plane surfaces. Deflation wind erosion is usually greater than accumulations, except in situations where there is an available reserve of fine, ancient or current debris lying upwind that can supply material for the construction of dunes (periglacial deposits, for example, in cold deserts or high mountains, river alluvium in transit in large valleys, or beaches along the coast). Runoff events occur at random, like the rains from which they are born. When they do occur, they follow the old, abandoned channels and spread out into ephemeral sheets across the glacis and pediments, preserved by a moribund morphology. However, in the presence of loose rocks and steep slopes, these streams can also provoke catastrophic landslides or mudslides (in the Peruvian Andes, for example). Wherever vegetation exists, it is sporadic and has adapted to extremely dry conditions. There are no trees and the vegetation consists of isolated stretches of grasses (*Panicum*, *Aristida*) or climbing plants with water reserves (*Cucurbitaceae*) on sandy slabs, and bushes (that are often halophilic) (*Scirpus*, *Salsola*) on clayey stretches.

The aridity indices are all extremely high, but are only theoretical values, given the irregularity of precipitation: a Capot-Rey index value < 0.30; Gaussen index value > 360; UNESCO index value < 0.03 (Table 3.7).

Station	Capot-Rey index	Gaussen index
In Salah	0.25	360
Aoulef	0.15	360
Sebha	0.01	360

Table 3.7. *Values for the aridity indices for hyper-arid regions*

This kind of a hyper-arid desert environment can be found in the heart of the central Saharan region, in the Touat and Tidikelt basins (Reggane, Aoulef), Fezzan (Sebha), Tanezrouft and Ténéré (Bilma). It can also be found in the Libyan-Egyptian desert, around the Red Sea, in Arabia (Nefoud and Rub al Khali), in the Iranian Lut desert, in the Chinese Taklamakan desert, in the United States (Death Valley), in Mexico (at the bottom of the Gulf of California) and along the coasts of Chile and the Namib.

3.4.8. *Conclusion*

This overview of bioclimatic classification, as with any classification, does contain some amount of subjectivity (the choice of criteria) and arbitrariness (the choice of thresholds and limits). While valid for the central and western Saharan regions, and while it *should* apply everywhere, the classification does not cover all possible situations in all the deserts of the world. Indeed, it might even be risky to use them in all situations. Given local variations in rainfall patterns, vegetation, and relief, as well as the unequal density of measurement stations, any extrapolation of intermittent observations over an entire region must always be done with caution. With regard to the scale, the same scheme cannot be applied to a slope and to an entire region, and with regard to environment, not all criteria have the same relative value everywhere. There is also the question of limits: there are enclaves, extensions, islands; on the maps that are drawn up based on the criteria used by naturalists, for example, the least arid facies plunge into the most arid facies following talwegs that form increasingly narrow digitations. Analysis through remote sensing is a most important contribution here. There are also many other nuances, which are at least as important, that appear to be functions of the latitude (hot deserts, cold deserts), continentality (coastal deserts, inland deserts), altitude (deserts along the plains, deserts in the mountains), the nature of soil (stony deserts, sandy deserts), topography (flat deserts, rugged deserts), and those that are functions of the morphogenetic agents or the geomorphological history of the region.

Whatever the reason, there are always perceptible nuances in almost all deserts that extend over a certain area on the Earth's surface. Although these classifications are often imperfect and arbitrary, they do, at least, offer a reference framework, for comparison and reflection into which we can introduce the other regional elements that make up desert landscapes.

3.5. Variations in climatic aridity over time

Climatic aridity varies over space and also over time. These variations occur over different durations, with a variety of consequences on the geodynamic evolution of desert regions and the constraints these regions impose on those who live here.

3.5.1. *Variations on a geological scale*

As with the other bioclimatic elements, over the course of the Earth's history deserts have undergone various changes in appearance as well as location. Some deserts have disappeared altogether, others have moved, new deserts have been born, while yet others have persisted for over thousands, or even millions, of years.

3.5.1.1. *Means of studying deserts*

It is not always easy to identify deserts from the past. The study of these deserts is based on a variety of fossil evidence, which are, unfortunately, very rarely brought together. Further, the analysis could be global or zonal, with regional diversities hard to determine in most cases. The most common kinds of evidence, which are generally the most reliable, are of a sedimentological nature. Applying the actualism principle, according to which the present holds the key to the past (Lombard 1972; Miskovski 1988), requires looking for similarities between ancient facies and modern-day facies: the nature of the composite material, sedimentation forms and characteristic deposits. A large proportion of detritus from cold or hot mechanical fragmentation (frost-weathering and/or thermal stress weathering) is considered a sign of aridity, while products of chemical alteration suggest a border climate with a dry season: a temperate sub-arid system, as in the Maghreb, with limestone or siliceous crusts, or tropical heat, as in the Sahel, with red oxidation (hematite) and ferricrete (duricrust). Evidence of wind action is also widely used as a criterion for identifying climatic aridity: particle size analysis of the transported material (sand or silt), whether loose or ultimately consolidated into sandstone ("ancient sand", as Monod described it); shaping linked to the transit, as with "rounded" quartz grains tarnished by weathering, polished or mottled windblown stones (ventifacts), stones with three faces (*dreikanters*), oblique eolian stratification or cross-stratification and lines on the surface of rocks (*windmarks*). Aridity can also be deduced in the presence of evaporites: salt, gypsum and salt deposits in chotts or sebkhas. Other evidence occurs more irregularly. It is sometimes possible to identify frequent shapes in the arid model: traces of corrasion, glacis and pediments, inselbergs and roots of dune systems. In some cases, it may also be possible to find fossils (Toumaï): bones, coprolites and petrified wood, although, by definition, these are rarely found in the desert.

Taken in isolation, these pieces of evidence are not enough to conclusively declare a region as being arid. We must also take into account the fragility, alteration or even destruction of significant pieces, as well as the possibility that certain pieces could have been imported from outside, and the fact that certain natural phenomena are ubiquitous (for instance, the wind is not a feature exclusive to deserts). Bringing together various strands of convergent facts and building them up into a system, and the generalization of these facts over a region, remains an essential step in supporting this interpretation.

Modern techniques, which are constantly being refined, can, however, make it possible to better analyze, specify, date and even quantify the data gathered, thus going further in paleoclimatic reconstitution. For example, from samples taken in continents or from marine sediments, it has been possible to determine the proportion of atmospheric gas trapped as inclusions in rocks when they were formed, especially CO_2, a greenhouse gas. These analyses have thus made it possible to draw up a table of the climates that succeeded one another through geological history and to compare this with what we know from the relative position of the continents, with respect to the poles and the equator, at present. Finally, various paleomagnetic methods and absolute radiometric dating techniques have helped position all these events in the chain of historical events going back from the present (BP = Before Present; 1950).

The second half of the 20th Century thus saw the development of more scientific paleoclimatology (Lamb 1977; Joussaume 1999; Nesme-Ribes and Thuillier 2000), which has, of course, proven to be more reliable for recent eras (the Tertiary and, especially, the Quaternary periods) than for older periods, and more reliable for the ice ages, which left behind many specific deposits, than for the more "discreet" deserts.

3.5.1.2. *The deserts of the past*

Using these methods, paleoclimatologists (Lamb 1977; Frakes 1979; Tardy 1986) and geologists (Aubouin 1975; Termier and Termier 1979) have been able to detect the existence of deserts, or at least the existence of aridity, through all ages, although caution is advised when studying those that are farthest removed from our times. In the early Precambrian period (the Archean Eon, dating back to beyond 2,500 Ma), any indicators were masked by a generalized metamorphism. On the other hand, red sandstone facies (assumed to be sub-arid), and the first evaporites appeared during the Proterozoic Era (2,500–1,000 Ma) and the Infracambrian period (1,000–540 Ma), notably in North America. The average temperature was high (a greenhouse effect due to a high proportion of CO_2) and the atmosphere was increasingly oxidizing (photosynthesis began in this period with the proliferation of blue algae.) In the absence of animal or plant life on land, the Precambrian continents were deserts in the *biological* sense, at least.

Paleozoic

From the continental Paleozoic Era (540–250 Ma) we know that there were two large glaciations, each following a significant orogenic cycle: the late Ordovician glaciation (around 450 Ma), which took place in the Sahara, following the Caledonian orogeny, and the late-Carboniferous glaciation (around 300 Ma), which occurred after the Hercynian orogeny. But what about deserts?

The existence of frankly arid evaporites (salt and gypsum) has only been observed around the Lower Paleozoic Era (540–400 Ma), dating to the Cambrian period in Siberia and the Silurian period in North America (modern-day Michigan). While it is not always possible to identify whether these are the result of hot aridity or cold aridity, it is certain that several major events contributed to diversifying landscapes: the creation of elevated relief tended to reduce temperatures, at least at altitude, and set off very active erosion processes. At the same time, living things started taking over *terra firma*.

Sequences of thick layers of detritic material (up to several thousand meters thick) of the sub-desert type appeared in the early Upper Paleozoic Era, in the Devonian period (400–355 Ma). The most famous of this is the *Old Red Sandstone* sequence, originating from the demolition of the Caledonian ranges. This is quite a complex set of detritic material that includes conglomerates deposited during floods or sediment from streams; sand shaped into cross-stratified dunes by the winds, and later consolidated into sandstone; and silty shales and evaporites that are evidence of lakes or sebkhas, sometimes containing crustaceans or primitive fish. The deposits are chiefly found in Ireland, Wales, Canada and Scandinavia, all of which formed a single North Atlantic continent, which was separated by a sea (the Ural Ocean) from the Siberian continent (also called Angaraland). Another sea isolated a Nigritian continent in the Southern Hemisphere, which was still cold, and which was made up of Brazil, Africa, Arabia, Madagascar, India, Australia and Antarctica.

In the Carboniferous Era (355–295 Ma), the Hercynian ranges were formed through the collision of continents coming together to form a single continent, Pangea, with the equator passing through from California to North Africa. In the northern part of this continent, North America, the Maghreb, Europe and Asia, this period was the apogee of coal-forming vegetation, which was considered to be tropical and which consumed large amounts of carbon. In the southern part of Pangea, with the South Pole at this time being located between Africa and Antarctica, there was a large permo-carboniferous glaciation that developed, the traces of which can still be found in South Africa, India and Australia in the present. However, there is very little information on deserts of the Carboniferous Era, as the orogenesis broke up the terranes. However, we do know that there was a zone of (possibly subtropical?) evaporites dating to the upper Carboniferous Era, formed to the north of the coal forests in North America, Greenland, septentrional Europe, and as far as Siberia.

In the Permian period (295–250 Ma), this zone of evaporites slowly shifted southwards. New sub-arid detritic sequences in the form of conglomerates, sandstone and shales appeared when the Hercynian ranges broke up. These were the *New Red Sandstone* sequences, with flora that lived in warm environments, found in Germany, the Vosges, England and North America (the Colorado canyon). Sequences of the same type, but with "cold" flora, existed in this same period in Siberia and in the Southern Hemisphere, on the continent of Gondwana (which was the successor to the Nigritian continent), notably in India and southern Africa.

The Mesozoic Era

The Secondary or Mesozoic Era (250–65 Ma) was a period that saw great developments in land-based life and the rapid shrinking of biological deserts. This period was also marked by the breaking up of the continent of Pangea and the start of a large-scale "continental drift" (Wegener 1915; Allègre 1983; Tardy and Roquin 1998) and the slow migration of the poles and the equator toward their present locations. The relative movement of continents, with respect to the poles and the equator, had several important consequences. The large bioclimatic zones that lay parallel to the equator slowly moved southwards, the glaciations disappeared, a continental climate that was overall warmer and drier emerged across most of the landmass, and an area of climatic aridity was set up between the tropical environment in the lower latitudes and the temperate environment in the middle latitudes.

The Triassic period (250–200 Ma), in many respects, maintained the conditions of the upper Permian period, and was generally very warm and very dry. It was also the richest in terms of evaporite deposits. The sub-arid New Red Sandstone, often including saline formations, continued to accumulate in Laurasia (which consisted of septentrional America, Europe and Asia) in the northern part of Pangea, in the regions that are now Germany, France, England and North America, fossilizing the post-Hercynian erosion surfaces that were being formed. In the south, the continent of Gondwana (Brazil, Africa, Madagascar, India, Australia and the Antarctic), separated from Laurasia by a mesogean sea that would become the Tethys Ocean, in turn saw the emergence of deserts. Red formations that were comparable to those in the north (sandstones, shales and evaporites), but which had formed in a colder environment (the flora being of the genus *Glossopteris*), succeeded the permo-carboniferous glacial formations in India, Madagascar and southern Africa.

In the Jurassic period (200–135 Ma), continental drift was accelerated and its climatic consequences became clearer. Gondwana moved toward the north, tilting westwards and this brought it close to the equator, which at the time passed through the Sahara. At the same time, India and Australia-Antarctica detached from Africa and independently began moving toward the northeast, moving further away from the South Pole. The climate tended to grow warmer on all these continents. Laurasia, in the meanwhile, began a clockwise rotation such that its western part moved closer to the North Pole, while its eastern part moved away from the North Pole. The climate grew cooler in the west and warmer in the east. A temperate zone appeared in the north, while a subtropical or desert zone appeared in the south. Arid facies disappeared from Western Europe and North America and appeared in North Africa (the Algerian-Moroccan Sahara) and the southwest United Sates (Arizona) in the low, post-Hercynian peneplain lands. In these semi-desert regions with dunes and evaporites, located on the edge of forests and swamps populated by giant conifers (*Araucariaceae*), large dinosaurs and primitive birds, torrential floods brought in vast alluvial deposits, made up of gravel and quartz-pebble sand, the bones of vertebrates, fish teeth and tree-trunks that would become the petrified wood found in the Sahara and American deserts. It must be noted that the Jurassic period, which was considered a period of orogenic calm, was also the period when oceans expanded over continental regions, starting with the Liassic transgression (200– 175 Ma). The global morphological and climatic trend was to produce greater homogeneity in warm environments, as in the Triassic period, but with more humidity, as in the Cretaceous period.

This continental drift continued into the Cretaceous period (135 to 65 Ma). The division of Africa and South America, which had begun in the Jurassic period, moved northwards. The Atlantic Ocean opened up, but Northwest Europe remained attached to Greenland and North America. The Indian Ocean formed between India, Madagascar and Australia. Sub-arid continental facies still endured in North America, the Sahara and southern Africa until at least the beginning of the Middle Cretaceous period (Albian, 108–96 Ma). Sandstone (generally red sandstone, locally rich in quartz), sometimes accompanied by evaporites or clayey silt and containing dinosaur remains and petrified wood, buried the last vestiges of the Hercynian relief: this was the "continental intercalaire" in the Sahara, wedged between the marine Namurian and Cenomanian strata. The re-forming of the ocean floors that followed the displacement of continental plates is likely to have been one of the causes for the large Cenomanian-Turonian boundary event, which, in the middle Cretaceous period (96–90 Ma), covered a significant part of the northern African continent, the Near East (as far as Turkestan), as well as the southern part of the North American continent (Texas, Mexico). In this period, the low and arid zones in the Sahara were underwater, leading to the deposit of limestone beds that today form many of the hamadas in the desert.

During the entire Cretaceous period, the climate on all continents remained warm (it is estimated that it was 6–10°C warmer than present-day temperatures) and tropical flora and fauna existed as far as Mongolia. However, with the expansion of the seas and oceans, conditions gradually became more humid and thus were less conducive to the expansion of deserts. Although the signs of a new orogenic cycle appeared in the late Jurassic period, notably in the Andes, it was in the Cretaceous period, following the compression of the sediment from the Tethys Ocean between Africa and Europe, that the first "alpine" formations appeared, in the Alps (as we know them today) and in North Africa. The terminal Cretaceous event (90–65 Ma) was a period that saw the seas retreating. It was also a period of intense eruptions (in the Deccan traps, Ethiopia and Brazil). Large quantities of gases and volcanic dust were released into the atmosphere and added to this were gases and dust produced following a giant meteorite that fell close to the Yucatan peninsula. The sky was completely covered and this caused global temperatures to fall sharply, leading to the extinction of the dinosaurs and several other large reptiles, who succumbed to climate change, disruption of their food chain and increased competition from mammalian fauna. Along the continents, which once again grew wider and developed relief features, rain shadow deserts appeared at the heart of and along the edges of massifs, in conjunction with localized deserts, which were expanding.

The Cenozoic Era

During the Tertiary or Cenozoic Era (65–2 Ma), the poles approached their current position, continental plates settled into the positions that they occupy today and the large oceans grew clearly distinct. North America separated from Europe and was only connected to Asia through the Bering isthmus, which would episodically emerge or be submerged. The Alps and the mountains of Iran emerged from the Tethys Ocean, which separated the African plate from Eurasia. India drifted northward from Madagascar and Australia and collided with the continent of Asia, creating the Himalayas. The American plates continued to move westward, leading to the creation of the coastal mountain ranges in the north, and the Andes. This new arrangement of continents led to new oceanographic, climatic and biological conditions, which foreshadowed present-day conditions. The opening up of oceans led to exchanges along the meridians, contributing to the differentiation of climatic zones along latitudes (a feature that had already become noticeable in the Cretaceous period). The large desert zones occupied a space that was adjacent to their present location. The localized partitions created by new relief features led to new rain shadow deserts. Terrestrial fauna was completely renewed with the rise of mammals. Species could travel more or less easily between all continents, except in the territories that had been isolated by the breaking up of Gondwana (Madagascar, Australia, Antarctica) where unique, endemic species developed.

The cooling that had occurred at the end of the Cretaceous period lasted a relatively short period, perhaps 10 million years. From the beginning of the Paleogene period (65–34 Ma) and throughout the Paleocene period (65–53 Ma), the average temperature once again started rising. The cause cited for this is the greenhouse effect created by CO_2 and aerosols thrown up by volcanoes, especially along the mid-Atlantic Ridge and Pacific Belt. The maximum temperatures were attained during the Eocene period (53–34 Ma). At this point, fauna that are found in warm or temperate environments still occupied Greenland. Arid or sub-arid facies (conglomerates, sand, silt, evaporites, riparian formations) existed on either side of the equator, between 20° and 40° North and South, and as far as 50°N in America, Europe and Asia. Off the Saharan coast, marine sediments, rich in very fine sand, show that the wind, although not very violent, was still present and effective. Toward the end of the Eocene period (toward 35 Ma), the Alps were practically in place and the North Atlantic had definitively opened into the Arctic basin. The continents and deserts now occupied the configuration that they do even today. The global temperature curve began to grow smaller and more jagged and has continued to do so up to the present day.

The Neogene period (34–2 Ma) began and ended with two extinction events, whose limits vary depending on the author and the criteria they use (sedimentological, paleontological or climatic). From the Oligocene epoch (34–23 Ma), temperature fell suddenly and then slowly rose again, especially in the lower latitudes, before finally falling. In the Southern Hemisphere, during the Miocene epoch (23–5 Ma), the continent of Antarctica, in which the South Pole is located, was isolated, by Australia moving further away and by losing contact with South America. Australia reached those latitudes where, as in southern Africa, deserts were created. The Drake passage opened up between Antarctica and South America, removing the last obstacle between the Southern Ocean and the Pacific Ocean. A circumpolar current was set up and the ocean temperature dropped by 1–2°C on the surface and 4–5°C in the deep ocean. From 30 to 20 Ma, sea ice appeared, and around 15–10 Ma an ice sheet formed over the continent. From this time onwards, there were exchanges between the neighboring oceans: the cold water in the depths moved northwards under warm, tropical waters along the coasts of Africa (the Benguela current) and South America (Humboldt current), playing an essential role in the locations of the coastal deserts. In the Northern Hemisphere, on the other hand, the Arctic Ocean remained relatively cut-off by the Bering shelf and the pathway of underwater ridges that link Greenland to Scotland. Cold water from the depths could only move beyond this region with difficulty and the warm currents that reached this far (currents from the Gulf Stream and the Kuroshio) maintained a temperate climate and subtropical flora as far as the middle latitudes. The Miocene was also marked by the paroxysm of the "alpine" orogenesis from the Mediterranean to the Himalayas. The disappearance of the sound to the east of the Urals reinforced the continentality of the Eurasian region. The Rifan-Andalus arc that connected Africa with Spain closed off the sea to the west, while Arabia, which had started to detach from Africa following the opening of the Red Sea rift, joined up with the Asian continent and lost its link to the Indian Ocean. By the end of the Miocene epoch (Messinian, 7–5 Ma), the former Tethys Ocean had transformed into a string of inland seas and residual lakes (the Mediterranean Sea to the west, the Black Sea, Caspian Sea and Aral Sea to the east), which contained evaporite deposits (gypsum, in particular). At the same time, large deposits of semi-arid (Pontian) continental sedimentation were accumulating.

This sedimentation continued during the Pliocene epoch (5–2 Ma) to such an extent that, in the continental environment, it is sometimes hard to distinguish it from the Miocene epoch. Thus, "Mio-Pliocene" sequences (with no differentiation between the eras) were observed chiefly in the arid regions. In North America and Asia, these filled intramountainous basins. In Europe, these were red formations and evaporites in Spain, Italy and the Balkans. In the Sahara, they formed the "continental terminal", essentially made up of conglomerates, more clayey or less clayey soil, evaporites and riparian deposits, especially thick in the foothills of the Atlas Mountains, where they support the hamadas found higher up. From a climatic point of view, the Pliocene was colder than the Miocene but warmer than present-day conditions. All the large relief features and climatic zones, including the deserts, were in place by this epoch and world geography had taken on its present appearance.

3.5.2. *Climatic changes and deserts in the Quaternary period*

At the end of the Pliocene epoch, significant changes presaged a new chapter in the Earth's history. First of all, global temperatures continued to fall. Fauna that lived in cold environments invaded temperate seas like the Mediterranean. On land, the first glaciers in the Northern Hemisphere formed, in the mountains of Alaska and Europe, around 2.7 Ma. Two glaciations (Biber and Donau), separated by a warmer interval, occurred in the period between 2.7–1.6 Ma. Some mammals from the Tertiary age, such as the Hipparions and the Mastodons, died out between 2 and 1.3 Ma, giving way to new genera of animals of the steppes and savannas, such as the elephant (*Elephas*), buffalo (*Bos*), horse (*Equus*) and camel (*Camelus*). Among the mammals of the order Primate, alongside the Australopithecus, that had existed since 5 or 6 Ma, there now appeared the earliest specimens of the genus *Homo*: *Homo habilis* around 2 Ma, and then *Homo erectus* around 1.8 Ma (see Chapter 2). In stratigraphy, this transition period is called the Villafranchian age (from Villafranca d'Asti, in Italy), the continental equivalent of the Calabrian age for the seas. Geologists had initially considered the Villafranchian age to be part of the Tertiary, and the last division of the Cenozoic Era. However, its novel features finally led the International Geology Congress in London (1948) to place it in the Quaternary, the most recent geological period (Flint 1971). The division between the Pliocene and the Quaternary, therefore, lies somewhere between 2.7 and 1.6 Ma. For greater precision, we generally turn to paleomagnetism: the Quaternary is likely to have started with the Olduvai event (when the magnetic fields were oriented in the same direction as today) within the Matuyama "reversed" geomagnetic epoch, around 1.8 Ma.

3.5.2.1. *Knowledge of deserts of the Quaternary period*

From this date onwards, the history of the Quaternary period (Wilson, Drury, and Chapman 2000) is marked by two main features: the development of humanity and the alternating of cold periods (called glacial) and warmer (or less cold) periods, called interglacials, accompanied by the regression and transgression of seas, and phases of sedimentation and erosion on continents. It is conventionally considered that there were four such alternating cycles, clearly defined in the Alps, the plains of septentrional Europe and North America, not taking into account the oscillations or "stages" within each cycle (Table 3.8).

Dating back thousands of years (ka) BP, these were, consecutively: the Günz glaciation = Nebraska (1,200–700 ka), the Mindel glaciation = Kansas (650–350 ka), the Riss glaciation = Saale = Illinois (300–130 ka) and the Würm glaciation = Weichsel = Wisconsin (80–10 ka), separated by interglacials. Based on this, the Quaternary was further subdivided into two subsystems: the Pleistocene, which included all the glacial cycles, and the Holocene or Postglacial epoch, which led to the current age. These divisions, which are a bit theoretical, offer a useful chronological time frame, but also require some comments.

Age (ka BP)	Periods	Glaciations		Prehistory		Climate
		Alps	America	*Homo*	Industries	
					Neolithic	
	Holocene	Postglacial				Interglacial
					Mesolithic	*Chadian*
10						
20						*Ogolian*
	Higher	Würm		*Homo sapiens*	Aterian	*Soltanian*
40		=	Wisconsin	*sapiens*		*Saourian*
						Glacial
70	Pleistocene	Weichselian		*Homo sapiens*		
80				*neandertalensis*	Mousterian	
						Tensiftian
						Ougartian
						Eemian
						Interglacial
130						
		Riss				Glacial
		=	Illinois		Acheulean	
300		Saalian		*Homo erectus*		
	Middle					Interglacial
350						
						Glacial
	Pleistocene					
650						
						Interglacial
		Mindel	Kansas			
700						
					Lower	
					Paleolithic	Glacial
1200	Lower					
	Pleistocene				Pebble culture	
1600		Günz	Nebraska			
1800						
	Pliocene			*Homo habilis*		*Villafranchian*
						Continental
		Donau				*terminal*

Table 3.8. *The chronological framework of deserts in the Quaternary period*

In the field, the episodes are not always distinct. For example, at the beginning of the Quaternary period, there was often a continuity of sedimentation that did not permit a clear division between this and the Pliocene epoch. Thus, in the language of geologists working on the Sahara, the peak of the "continental terminal", a large part of which is riparian, is often said to be "Plio-Villafranchian", "Plio-Pleistocene" or "Plio-Quaternary". This signifies that the arid or semi-arid conditions observed during the Pliocene epoch extended at least as far as the time of the Donau glaciation, toward 1,600 ka. The clearest division visible is the sinking of the hydrographic network and the creation of the earliest and highest fluvial terraces. However, this was also the result of Villafranchian tectonic activity. Consequently, the geodynamic variations depended on the climatic alternation between glacial/interglacial periods at high latitudes, each of which disrupted the general oceanic and atmospheric circulation (Leroux 1996). In the deserts, this was manifested as a simultaneous variation in pluviometry, in the form of "pluvial" and "interpluvial" periods. At the same time, geomorphic processes created a unique environment, composed of relief forms and correlative wet or dry formations, which were sometimes juxtaposed with one another, sometimes layered and sometimes nested (Arnaud-Fassetta 2008).

These Quaternary geodynamic episodes, which succeeded one another over a very short timespan (<2 Ma) left many well-preserved features, which were easier to interpret than the older findings. In addition to the usual geochemical evidence (evaporites, alterations, paleosols, crusts, duricrusts, etc.) and the sedimentological evidence (fluviatile wind-borne or riparian sediments), geomorphic data were also available (talwegs, terraces, glacis, open plains, lakes, dunes, ergs, sebkhas, etc.), which revealed transformations in the landscape. The nesting of forms within each other and the superimposition (or layering) of formations also made it possible to establish a relative chronology of events and to retrace the chain of climate systems that produced them. However, chronological correlations are not always exact from one region to another. There may be a difference of several thousands of years between the datation carried out using the fluvio-glacial terraces (in the Alps) and the datation carried out in the moraines (Europe and North America) or in the desert landscapes. The succession of flora that made up the vegetation (represented by pollen conserved in riparian sediments, swamps or bogs) is also a good indicator of the paleoclimatic variations in the environment (after, of course, excluding any foreign particles that may have come with the wind). It is also possible to get valuable data on local conditions from studying reptile and micromammal fossils, while the study of large animals, who are very mobile, is more helpful in providing information on biomes. The presence of humans was also a new and rich source of information and reflection. Through their bodily remains, industries, artistic representations (cave paintings), religious ceremonies (funerary rites) and social practices (pastoral or agricultural practices, various constructions and, from the 3rd millennium BCE, writing), prehistoric humans, in turn, became a source of knowledge on the environment and an instrument to measure time. However, human industries, which are by-products of civilization, did not appear simultaneously worldwide and, therefore, could not serve as irrefutable proof of contemporaneity. Despite this, however, for the most recent periods of the Earth's history (Holocene), human remains and artifacts have been highly sought after.

In the second half of the 20th Century, the Deep-Sea Drilling Project, as well as drillings carried out in lakes and in the vast polar ice sheets (Greenland, Antarctica), provided a wealth of data that soon became primordial. The marine sediments that were recovered by drilling along the borders of deserts (in the Atlantic Ocean, off the Saharan coast, for instance; or in the Indian Ocean, off the continent of Africa, or in the Southern Ocean, to the south of Australia) have provided information on the climatic conditions at the time that they were deposited through their lithological composition and the microfossils they contained. Like the pollen, dusts, especially volcanic dusts, trapped in ice and sediments make it possible to establish the provenance, force and direction of the winds that carried them. It is also possible to evaluate the ambient temperature, using heavy isotopes in the water ($^{18}O/^{16}O$ and ^{2}H), the quantity of which varies depending on evaporation (which itself depends on temperature). As concerns absolute radiometric datation, the short timespan under consideration makes it essential to use radioactive isotopes with a short half-life. The most commonly used isotope is "Carbon-14", (^{14}C), which is applied to carbon-rich samples, such as limestone shells, organic debris and wood or coal. However, this technique does not make it possible for us to date objects older than 35,000 years (some researchers say 40,000 years). For older specimens, there are tools such as thermoluminescence (for between 35,000 and 100,000 years), the analysis of the uranium/thorium ratio ($^{238}U/^{230}Th$) (for between 70,000 and 300,000 years), and the potassium/argon ratio ($^{40}K/^{40}Ar$) for datation of objects over a million years old. Thus, an excavation such as the Vostok excavation in Antarctica between 1970 and 1998, carried out over 3,600 m, offered a complete cross-section of the Quaternary, spanning 420,000 years (Petit *et al.* 1999). The problem that arises is that the proposed dates vary from one author to another, with discrepancies that could be as high as several thousand years, for the oldest specimens, and up to several hundred years for the newest, depending on the methods used. Other uncertainties stem from the density of the network of observations, because when the objects to be dated are rare, widely scattered or unequally distributed, there are large gaps between them in both space and time.

The alternating climatic conditions that marked the whole of the Quaternary period are only somewhat related to continental drift or orogenic deformation, which are almost imperceptible over such a short timespan. The primary reasons for the alternating conditions are extraterrestrial, as theorized and demonstrated by the Serbian astronomer M. Milankovitch, in the 1920s and 1930s (in (Köppen and Geiger 1930–1939)). The conditions were related to cyclical variations in the orbital parameters (the "orbital forcing") of the Earth's movement around the Sun. M.

Milankovitch calculated a period of around 100,000 years for the eccentricity of the orbit and 40,000 years for the polar axis to tilt to the ecliptic axis. We find similar results in the periods, stages and oscillations of the large glacial and interglacial periods. These variations affect the amount of sunlight received and, hence, the amount of energy received, the temperature, evaporation, humidity in the air and, in deserts, the volume of precipitation received. In the history of the Quaternary period, the most recent episodes are the most well-known, especially those that occurred in the last 130,000 years, spanning the Riss-Würm interglacial (130 ka–80 ka), the Würm glaciation (80 ka–10 ka) and the Holocene epoch (>−10 ka). International research groups like CLIMAP (Climate Long-range Investigations, Mapping and Prediction) are studying this on a global scale. In the context of deserts, the paleoclimatic conditions (Petit-Maire 1984; Rognon 1996) and paleometeorological conditions (Leroux 1996) made it possible to broadly identify major divisions that contained one or more warm and humid episodes of varying length, which alternated with more arid and colder periods:

– the *last great interglacial*, the Riss-Würm interglacial, began around 130 ka with "a climatic and hydrological optimum" (Eemian *s.s.*). Another climatic optimum occurred during the final period of the Pleistocene warming, between 90 and 80 ka. Intermediate dry episodes occurred around 116 ka–98 ka. The interglacial ended around 80 ka, when the region was affected by the progressive effects of a cooling period that began in the higher latitudes;

– the increasing or Anaglacial phase of the Würm ice age first touched Greenland and then North America (present-day Wisconsin). The Arctic sheets led to icebergs appearing in the Atlantic. A cold phase settled over the entire Northern Hemisphere. All marine excavations indicated that there was extreme aridity between 70 and 50 ka. Unfortunately, continental evidence is still mired in uncertainty regarding the chronology, given that ^{14}C ceases to be reliable around 35 ka, while U/Th datation is only effective from about 70 ka. On the other hand, it is certain that there was a humid and cold pluvial episode between 40 and 20 ka, with a recurrence of aridity around 30 ka. This was followed (between 22/20 and 13/9 ka) by a new arid period that corresponded to the Last Glacial Maximum (the formation of the Scandinavian ice sheets and ice sheets in the Alps; sea ice covered the Atlantic as far as the 45th parallel). An opposite phase set in around 13–12 ka, during the decreasing or Cataglacial phase in the Würm ice age, with a progressive pluviometric shift toward the north. This period marked the transition from the Upper Pleistocene to the Holocene, generally placed around 10 ka;

– the return to pluvial conditions reached its maximum around 9–7 ka, during what is called the Holocene climatic optimum. This was a warming period that spread from south to north and from southeast to northeast as the Euro-American ice sheets retreated, with summer temperatures in Europe reaching 2–4°C higher than today. This was also the period in which the Neolithic civilization flourished, reaching its apogee between the 7th and 5th millennia BCE, following the historical chronology. From this time onwards, aridification continued with periods of severe drought, or shorter periods of thaw, developing more or less quickly depending on the location.

3.5.2.2. *Pluvials and interpluvials*

Considering current meteorological conditions in the Sahara in winter, the first researchers who studied the history of the Quaternary period in North Africa (working in the 1950s) initially assumed that the *pluvials* in the Sahara corresponded, chronologically, with the **glacials** in Europe, and that the **interpluvials** corresponded with the *interglacials*. They had observed that the rainy season in septentrional Sahara coincided with the cold season in Europe, and that the dry season coincided with the hot season. The fact that, in the Quaternary period, there were peri-glacial formations in the Tibesti and glaciers in the Atlas and Hoggar ranges seemed to bear out this idea. However, their error lay in the fact that they had too quickly generalized this pattern (seen in the septentrional region) to the whole of the desert, failing to take into account the fact that in the south (Sahel), the rainy season occurred in the summer[4] and the dry season in the winter. Climatologists (Dubief 1951; Dubief 1959–1963) and prehistorians (Balout 1952) quickly recognized that the seasons were "flipped around" between the North and South. This then led to the idea that the pluvials in the north were likely to have corresponded with the interpluvials in the South and vice versa. In reality, however, the situation is more complex, as can be seen in Table 3.9. However, these facts could not have been deciphered without the many and varied investigations carried out in the field, using advanced techniques from the field of radiometric datation, paleobiogeography and paleometeorology.

The data from Table 3.9 show that each major change in global temperature that produced, notably, variations in the volume of polar ice was accompanied by a significant change in pluviometry, leading to a pluvial, in the case of overall warming (130 ka, 10 ka), or leading to an interpluvial, in the case of cooling (80 ka, 22 ka). However, the table also shows that, within each glacial or interglacial, there have always been shorter, alternating wet and dry periods that researchers have tried to relate to the small oscillations in the Milankovitch cycle. Indeed, each of these alternating periods produced a complete disruption in dynamic climatic factors: air temperatures, sea temperatures, ocean currents, humidity, atmospheric circulation and the position and intensity of centers of meteorological action (Leroux 1996).

4 In fact, because of this error, these rains were paradoxically called "the wintering".

Age (ka)	North Sahara supratropical	South Sahara and Sahel tropical	Age (ka)	Events
	Aridification			
5		Aridification		*Neolithic*
	Last pluvial Attenuated desert Sub-arid		6	
	Temporary lakes Swamps	Last pluvial		Post-glacial
7	Winter rainfall Temporary streams	?	7	
7.5	====Dry recurrence====	====Dry recurrence====	7.5	
	Sub-arid vegetation	Marine transgression *Nouakchottian*		Climatic
	200 km further to the south	"Green Sahara" Savannas and steppes		optimum
	Semi-permanent streams Dispersed lakes	"Great lakes Sahara" *Tchadian*		Holocene
	Winter temperate precipitations	Semi-permanent streams Large rivers		*Mesolithic*
9	Pluvial Holocene climatic optimum	Summer tropical precipitations (monsoon)		
	Generalized warming			
	Subdesertic vegetation Sebkhas Dunes	Pluvial Holocene climatic optimum	10	—
	Rare precipitations Temporary streams	Precocious pluviometric increase	13	
	Snow and glaciers in mountains	Sebkhas Large dunes *Ogolian*		Last
	Thermal Minimum at −18 ka	Arid vegetation Temporary streams		maximum
20	Very cold and dry arid	Desert progression of 500 km to the south		glacial
	Soltanian Saourian	Marine regression till −130 m		
	Winter precipitations	Cool and dry arid	22	
25	Cold and wet pluvial	Cool and wet pluvial humid	25	*Homo sapiens*
30	====Dry recurrence====	====Dry recurrence====	30	*sapiens*
				Aterian
	Cold and wet interstage	Cool and wet interstage		^{14}C
40			40	Würm glacial
				Homo sapiens
50			50	*neandertalensis*
	? Chronological uncertainty ?	? Chronological uncertainty ?		$^{16}O/^{18}O$
70	====Arid recurrence====	====Arid recurrence====	70	U/Th
80	Generalized cooling	Generalized cooling	80	
	Streams and floods activation	New extension of permanent lakes		*Mousterian*
	New climatic and	New climatic and		
90	hydrologic optimum	hydrologic optimum	90	
	Warm pluvial	Warm pluvial		
98	====Dry recurrence====	====Dry recurrence====	98	
100			100	
	Warm pluvial	Warm pluvial		Last great
110			110	interglacial
	Eolian formations Dunes	Eolian formations Dunes		
115	===Dry recurrence====	====Dry recurrence====	115	
120			120	
	Tensiftian Ougartian	Savannas and peridesertic steppes		*Acheulean*
	Sub-arid to sub-desertic steppes	Weathering Ferruginizations Cretes		
	Weathering gypsicretes calcretes	Semi-permanent streams Large rivers		
125	Semi-permanent streams Lakes	Very large lakes	125	Climatic
	Winter precipitations	Summer precipitations (monsoon)		optimum
	Climatic and hydrologic optimum	Climatic and hydrologic optimum		
130	Warm pluvial *Eemian*	High temperature pluvial	130	

Table 3.9. *Comparative scheme of pluvials and interpluvials in the Sahara*

It can thus be seen that pluvials and interpluvials cannot be exactly compared to glacials or interglacials, although both sets of phenomena have common astronomical or atmospheric causes. They are not entities of comparable scale. The glacials and interglacials represent major climate changes that correspond to correlated variations in global temperature and the volume of polar ice. The pluvials and interpluvials are more closely connected with local conditions of atmospheric circulation, which determine available humidity and rainfall patterns.

During the cold periods, evaporation, which was already limited by aridity, was reduced to its lowest point. Absolute humidity (see p. 68) was low, but, on the temperate side of tropical deserts, the low temperatures maintained a relative humidity that was sufficient to keep the sky frequently overcast. However, there was little or no rain, as there was no air rising (heavy air) except in the mountains. On the other hand, there was constant aerological activity, stimulated by the thermal contrasts between the poles and the tropics, and between continents and oceans. The movement of powerful drops of cold air originating in the polar regions (MPA), in conjunction with low-pressure areas that were, to greater and lesser extents, deep and mobile along the meridians, occasionally brought "Mediterranean-type" winter showers, with violent winds (tempests or sandstorms) as well as snow at high altitudes. These MPAs reinforced the stationary tropical anticyclones (AA) (see p. 71) that opposed the penetration of the monsoon. Dry, continental trade winds (harmattan) blow on the tropical slope and the eastern flank of these anticyclones. On the western front, there is a strong maritime trade wind that stimulates the upwelling of cold water from the depths that deepens the foggy aridity along the coast. The anticyclonic dominance creates dry, arid and interpluvial conditions everywhere, with reduced flow in rivers, a slow down in geomorphic activity (except for wind and frost) and smaller expanses of green cover. For instance, at the start of the Würm ice age (70–50 ka), this caused accentuated aridity, whose presence was revealed by the high quantity of eolian deposits in marine sediments off the coast of Western Sahara. Similarly, the Last Glacial Maximum triggered an arid interpluvial (22–13 ka) in the southern Sahara, this being the most arid interpluvial of the upper Pleistocene (Ogolian).

During the warm periods, on the other hand, the weakening of circulation along the meridians and of the MPAs led to a weakening of the tropical anticyclones, which were, themselves, often replaced by thermal depressions. Thus, there was nothing to prevent a large-scale penetration by the monsoon, which brought humidity and rains to the tropical and eastern fringes of the desert. Wind force was reduced and wind activity was less effective. The maritime trade winds were calmer and tempered the upwelling of cold water, making it less drying. At high temperatures, relative humidity is low, but evaporation, incoming humid air from the outside and frequent instances of air rising (light air) are all conducive to pluvial conditions, with renewed activities in the rivers, the recharge of groundwater tables, a profusion of lakes and the expansion of green cover. Thus, the periods of Cataglacial warming from 130–125 ka and from 13–5 ka were characterized by a hot and humid "climatic and hydrological optimum." Similarly, there was a softer climatic interstage in the middle of the Würm ice age (between 40 and 20 ka) resulting in a cold or cool pluvial.

These alternating periods correlated roughly to variations in global temperatures, but were deeply influenced by local geographic conditions such as latitude (zonal position), distance from the sea (continentality), ground movements (tectonics), location with respect to the large centers of action and axes of circulation in the atmosphere (meteorological environment), the arrangement of relief features (exposed, sheltered), altitude (bioclimatic zones) and so on. However, the alternating humid/pluvial and arid/interpluvial cycles were chiefly phenomena occurring on the edges of deserts. The temperate slopes were always cold in the winter and would have been very cold during the glacials. The region was dry, on the whole, ranging from sub-arid conditions to sub-desert aridity. Rainfall was rare and there were only intermittent flows. Even during the pluvials, the region remained a desert, but a "milder desert". The tropical side and the eastern edge were usually the most starkly different from other regions. They were less influenced by sub-polar disturbances and more sensitive to monsoons. The temperatures were always very high and the winds less violent. During the interpluvial, sheltered from the tropical anticyclone, there was severe dryness, the region was arid, a real desert, and was subject to wind dynamics exclusively. The pluvials, on the other hand, were very wet, with the formation of many lakes, semi-permanent rivers and sub-arid or even semi-arid vegetation, like that found in steppes or on the savannas. In the space between these two edges, there reigned an arid or hyper-arid region, alternately narrow and wide, where there was extreme and, above all, permanent drought.

Knowing that the climate changes of the Quaternary period depended primarily on cosmological causes, it is not surprising to find similar sequences in all the deserts of the world (Alsharhan *et al.* 1998), with the sequences being approximately synchronous with the glacial/interglacial cycles. There are, however, several differences in detail, with chronological gaps, which may be as large as several millennia, and which cannot only be attributed to the margin of error in the methods used or the low density of observation stations. There is also a slight delay, which is experienced unequally, between the orbital astronomical positions, the variations in the volume of ice, the temperature of the oceans (measured by the presence of ^{18}O) and of the continents, and the evolution of landscapes. The secondary oscillations (pluvials/interpluvials) are harder to identify and are also more sensitive to local geographic and aerological influences, as well as being the hardest events to synchronize.

3.5.3. *Variations on a historic scale – droughts*

3.5.3.1. *The concept of drought*

At the beginning of this chapter, we examined the difference between *aridity* (a concept applicable to space) and *drought*, which manifests in time. In both of these cases, there is a significant lack of water that hinders the development of life, along with a few other aggravating circumstances: excessive sunlight, high temperatures, high evaporation levels, dryness of air and soil, a dominant role played by the wind and desertification. While aridity is a permanent state of drought, a drought can also be considered temporary aridity, expressed with respect to a situation that is called "normal", i.e. the average situation over the longest possible period of observation. A drought is also not unique to arid regions but can occur in any region, irrespective of the climate. It is not a change in climate, but a complex climatic accident that is reversible and that also has pluviometric, hydrological, biological and even human aspects, and is defined by a large number of parameters with multiple serious consequences (Mainguet 1995).

A water deficit is not, in itself, an exceptional phenomenon. Indeed, it is normally included as part of the definition of the "dry season" that is common to several types of climates. However, it is a constraining factor which can turn into a drought when the duration of the water-deficit period is long enough and spread over a large enough space to cause a significant modification in the physical and biological environment. A water deficit is chiefly caused by a pluvial deficit, leading to a negative deviation from the mean. This compromises the humidification of the soil (edaphic or pedological dryness), affects the flow in springs and rivers (hydrological dryness) and prevents the recharge of groundwater reserves (phreatic dryness). It may also be caused or exacerbated by excessive evaporation. Drought is said to begin when a usual seasonal water scarcity extends beyond its normal duration into the preceding or following rainy season to stretch over the entire year or several years. This can also take the form of recurrent long, dry periods, interrupted by short intervals of rainfall. These humid intervals, however, are brief or limited and there is no intermediate recovery from the drought, whose harmful and cumulative effects prevail. A drought is chiefly perceived through its effects on the hydrosphere and biosphere, whether wild or modified for human use: a drastic shrinking of water resources, leading to low water in rivers and wells drying up, the degradation of green cover, grazing land and farmland, especially if the drought coincides with an important phase in the development of the vegetation like germination or maturation.

A drought has many effects, not all of which are felt equally, but are all the more devastating when combined with an environment that is already ecologically fragile, like deserts and the areas bordering deserts. From being a simple climatic aberration that is transient and sooner or later followed by a return to normalcy, a drought may then be truly catastrophic in demographic, economic and social terms, worse, even, than disasters such as earthquakes, volcanic eruptions or floods, as it is more enduring and more widespread than the other disasters (Lamarre and Pagney 1999).

3.5.3.2. *Significant droughts*

Droughts have always occurred, throughout time. After all, arid periods that spanned long durations, recognized in ancient periods, and which completely disrupted physical and biological conditions around the globe for thousands or even millions of years, may simply be considered droughts on the geological time scale. To continue with the Quaternary period, there is evidence of briefer and more numerous instances of drought, during the glacial periods as well as during the interglacials (Table 3.9). More details are available from the Holocene onwards, as a result of archaeological data as well as historical data. The large Würm–Weichsel–Wisconsin cataglacial deglaciation, which occurred through most of the Northern Hemisphere between 13 and 10 ka, put an end to the severe aridity that had set in during the Last Glacial Maximum. The Holocene climatic optimum, which then followed, was interrupted, in all zones, by a dry interval lasting between 500 years to a millennium, around 7.5 ka, a period long enough to decrease the volume of lakes, multiply evaporite deposits, build up dunes and disrupt flora and fauna. However, the major event of this epoch has been the long aridification that began between 6,000–5,000 years BP (the 4th and 3rd millennia BCE) and which continues into the present. Although there have been many fluctuations, in general precipitation has reduced and aridity is gaining ground.

These fluctuations include certain historic droughts. These were mentioned in the Bible, in literary texts and archives from the Mediterranean, India and China, all dating back to Classical Antiquity, and also in medieval Arab and Persian chronicles. Colonization introduced these records to Brazil and Mexico in the 16th Century, and to black Africa in the 17th and 19th Centuries. At this time, the climate in Europe was characterized by a temperature that was much lower than present-day conditions. The glaciers in the Alps as well as glaciers in Alaska and New Zealand extended far beyond their present-day limits. This was the "Little Ice Age", which occurred between 1450 and 1880, with an intensification between 1610 and 1860 (Le Roy Ladurie 1967). In Africa, there were long droughts lasting several years, 1681–1687, 1738–1756 and, most notably, 1828–1839, all of which occurred within this Little Ice Age and all of which caused famines (Nicholson 1978). In the late 19th Century and especially in the 20th Century, the development of observation networks and advances in data treatment techniques (spatial mapping, mathematical models, information processing) led to the discovery of many widespread droughts, nearly all across the globe (Sécheresse 1995).

At the end of the 19th Century, although a long drought was still claiming victims in Brazil (over 500,000 people between 1877 and 1879) and in Ethiopia (one third of the population and 95% of livestock perished between 1888 and 1892), the western Sahel was marked by an increase in precipitation between 1875 and 1895. After this period, aridity once again resumed its inexorable march across the region. Between 1895 and 1900, there was a long series of dry, or very dry years, interrupted by humid intervals, often localized rather than generalized.

An initial long drought occurred between 1896 and 1916, attaining its maximum between 1912 and 1915. During these years, the deserts were severely affected. The period from 1913 to 1915 was disastrous for the Sahel. Wells dried out completely and humans fled the drought-ravaged fields and grazing lands. It was the same story in Ethiopia. The drought seemed to touch almost every part of the world. In Mexico, there were 29 dry years between 1875 and 1910, 15 of which occurred largely in the arid north of the country; in Brazil, 1915 was one of the worst years in terms of water scarcity; in Central Asia, there was a drought from 1906 to 1908; in India, the summer monsoons failed in 1899 and again in 1918; and Australia's drought was between 1908 and 1910. There then followed a relatively humid period, although this was still marked by localized but devastating droughts: Brazil 1928–1932; Australia 1930, as harsh as the 1908 drought; East Africa, 1933–1934; Central Asia: 1938–1939.

A second widespread drought occurred in the middle of the 20th Century between 1940 and 1960: this affected the Sahel between 1940 and 1944, then again from 1948 to 1949; Mongolia was affected in 1944; Brazil between 1951 and 1955 and again, very severely, in 1958; Kenya and East Africa saw droughts in 1955 and 1965.

The last great wave of droughts began in the 1970s and lasted almost throughout the rest of the 20th Century. From 1970 to 1973, droughts hit West Africa, although this was a humid period in Ethiopia, the Near East and Brazil. Then in 1973–1974, it was East Africa that was ravaged by droughts as far as Tanzania. The droughts between 1983 and 1985 devastated both West Africa and East Africa. In fact, despite some instances of rainfall (limited and localized) the drought continued in most deserts until at least the year 2000, and its effects were often exacerbated by political crises and wars. There were, of course, certain years where the droughts were most intense: 1979 and 1983 in Brazil; 1984 in Kenya; 1985 in Chad; 1985 in Ethiopia, when famine caused over 800,000 deaths; 1998–1999 in the Near East; 1986–1987 and 1998–2001 in Central Asia and North China. In the year 2000 alone, severe drought once again affected the western United States, northern Mexico, the Sahel in East Africa, Iran, Afghanistan, Pakistan, Central Asia, Mongolia and North China, while there were humid periods in the deserts of the Southern Hemisphere and even dangerous floods in Northeast Brazil and eastern Australia.

3.5.3.3. *The great drought in the Sahel (the late 20th Century)*

The great drought that hit the Sahel in the early 1970s is often used as a model because it was methodically studied and analyzed (Mainguet 1995; Leroux 1996) and, being the first post-colonial drought suffered by the newly independent countries, it had long-lasting effects on the population and economy of an entire region. This region, the Sahel, is the "southern shore" of the Sahara, stretching from Mauritania to Mali, Burkina Faso, Niger and up to Chad around the 15th parallel North, which is a particularly fragile desert border, with an arid or sub-arid to semi-arid climate, with 100–500 mm of rain concentrated into a single, short season from July to September. This season grows even shorter as we move northward and has an interannual variability, which can lead to long gaps between rainfall, sometimes as large as twice the regional average between rainfall. The vegetation varies from desert plants to savanna vegetation, and even to forest cover. Nomadic pastoralists live in the north and subsistence farmers in the south. Although we know that droughts have occurred in the Sahel several times every century, the drought toward the end of the 20th Century was exceptional in the area that it affected, its intensity and its duration.

It covered practically all of West Africa and most of East Africa. After a long period of normal rainfall, between 1950 and 1967 (although the quantity did gradually decrease), 1968 marked the end of the rains and sounded an early warning; 1969 was an intermediate year with rains and then from 1970 onwards the rainfall deficit accelerated. A series of dry years followed in succession, growing increasingly drier (1970–1973; 1977–1978; 1983–1985) separated by years where the rainfall deficit was slightly smaller (1974–1975; 1979–1981; 1987–1988) but the humidity kept decreasing.

The drought first affected the most vulnerable points: flow of water in water bodies, the water levels in wells and permeable land, like the covered dunes from the Ogolian period. This created niches and vast expanses of drought-hit land, isolated from one another by large spaces that were still humid. The lack of water then affected groundwater reserves, leading to the drying up of springs, very low water levels in rivers, restricted irrigation and the exhaustion of reserves. The drought-affected islands thus multiplied, expanded and started joining up. Agriculture and grazing lands suffered and eventually the whole region was overtaken by the drought. Then signs of famine among humans appeared and the livestock started dying. At the same time, conditions were also exacerbated by human activity especially through people crowding into regions that had been less severely affected than others. These concentrations were ruinous for the steppe, which was ravaged by the excessively large herds of livestock, and also disastrous for trees, which were chopped down for firewood and to create open grazing land. Even the large colonial measures, which were supposed to help the population, proved to be harmful: the large tanks lost excessive water through evaporation and the high-flow artesian wells attracted such large populations of livestock that all lands were stripped of food for tens of kilometers around the water sources and the animals died not of thirst, but of hunger.

After 30 years of near-uninterrupted drought, further accentuated in 1997–1998, the rainy season in 2000 failed to meet the expectations of the farmers and pastoralists, especially in Niger, the Aïr and the Ténéré, as well as in Ethiopia and the Horn of Africa. Vegetation had declined drastically, the groundwater reserves had not been recharged, and goats had stripped the trees bare. The loss of livestock, whose populations had just begun climbing after the civil wars and past droughts, deprived the nomads of their last available resources and their life in the bush looked very precarious.

The drought in the Sahel highlighted the roles played by several convergent causes (Leroux 1996), which explained both the origins of the phenomenon as well as the reasons for its prolonged duration, showing it to be less extraordinary than it appeared. Certain researchers cited chiefly human causes, especially the overexploitation of vegetation through deforestation and overgrazing. The American meteorologist J. Charney (1975) – cited by Leroux 1996 and Joussaume 1999 – believed that these activities alone were enough to trigger the processes of desertification: the degradation of green cover increased the expanse of bare soil, which increased the albedo, decreased the amount of energy available in the lower layers of the atmosphere and prevented air from rising to form pluviogenetic conditions. All of this resulted in a pluviometric deficit and further deterioration in the green cover, and the process continued. The problem with this explanation is that the physical basis of this argument is debatable and human actions would not explain the droughts of the past. Indeed, when this phenomenon strikes, humans are more often the victim than the engine, although, of course, deliberate or accidental human actions may sometimes be a causative factor or aggravate existing conditions. The chief causes, however, seem to be natural and these are, in general, the same factors or causes that eventually cause aridity (see section 3.2, p. 67). In the case of a drought, these causes work over a given space and a limited period of time. Most researchers today agree that a phenomenon such as the great drought in the Sahel is most certainly a specific form of global variation in general atmospheric circulation, controlled by regional geodynamic factors (Leroux 1996).

3.5.3.4. *Droughts, climate crises and climate change*

The instances of drought around the world and their economic–political impact often provoke this question: are these always transient and reversible climate crises, or are we seeing a true global change in climate? There is no simple answer to this question. Climate change is an undeniable phenomenon (as has been statistically proven) and drought does lead to aridity (i.e. a permanent state) when the changes in the environment persist longer than a human life over several centuries or even millennia. Short term variations may mask the direction of long-term evolution. There is also the fact that what reliable data there are on temperature and precipitation is relatively recent and does not go back farther than 1860. Further, the majority of these data are only available for the Northern Hemisphere. Thus, the data gathered cover a very short period and is too localized to allow researchers to be certain about whether these are ordinary, repetitive fluctuations, or whether they are successive steps in a large-scale climatic transformation, aggravated, if not provoked, by human actions (Godard 2001).

For a long time, binary relations such as sunlight–temperature, temperature–precipitation, precipitation–aridity and so on were considered sufficient explanations for these variations. However, today it is recognized that there are more complex interactions with a system of interdependences that simultaneously involve the atmosphere, the oceans (in all three dimensions, as well as time) as well as the geodynamics of the surface and the biosphere (including humans). This, of course, does not make it any easier to solve the problem of why there are variations, and makes it even harder to attempt to predict future events. Climate change has thus become an important subject of scientific research from the late 20th Century onward. In the early 21st Century, the United Nations entrusted international and interdiscplinary bodies such as the World Meteorological Organization (WMO) and the Intergovernmental Panel on Climate Change (IPCC) with the responsibility of carrying out this research (Houghton *et al.* 1990, and following years).

The climate history of deserts in the Quaternary period show a certain correlation (or perhaps covariation) between changes in the average global temperature and changes in the paleoenvironments (Petit-Maire 1999, pp. 123–124): the cold period promotes aridity and humans forming more concentrated groups, whereas the warm period is more conducive to humidity and to humans moving further apart. During the Last Glacial Maximum (around 18 ka), the average global temperature was around 4.5°C lower than the present-day average. At this time, the Sahara was larger than it is today. It extended over the savanna at least as far as the 12th parallel, and with the seas retreating, the Atlantic coastline was about 200 km further west. On the other hand, Central Asia was largely covered by the periglacial tundra, but the deserts of East Asia, swept by loess-bearing continental winds, remained out of reach of the summer monsoon as they were far from the Pacific coast. A cold wave and drought disrupted the cataglacial warming that occurred in the Würm ice age, between 11 and 10 ka (the Younger Dryas) and accelerated the sedentarization of the hunter-gatherers of the Near East (Natufians) (Sanlaville 2000). During the Holocene climatic optimum, around 8 ka, the average temperature was about 2°C higher than today. The Sahara was green, the savanna stretched almost as far as the tropics, and all of Central Asia was covered by the steppes. Agriculture and animal-rearing practices began developing everywhere. From this time onwards, the global average temperature dropped once again and aridity progressed in a zig-zag manner. However, the coexistence that is observed today between probable global warming, the persistence of prolonged droughts and the progress of desertification seems to be contradictory.

The fact is that rather than the average global temperature, we must study thermal gradients: the gradients between the poles, which retain the cold, and the heat in the tropics; the gradients between surface temperatures of oceans and continents, and the effects of large currents such as the Gulf Stream or the El Niño phenomenon (see p. 78). These thermal inequalities, which are of a geographic order, produce continuous and complex heat exchanges through marine and aerological corridors. In more schematic terms, these exchanges produce the same effect as the changing of seasons (see section 3.2.1.5, p. 71) but over a longer time frame, through the displacement of an entire bioclimatic system toward the equator (in the case of cooling) or toward the pole (in the case of warming). On land, this displacement of the system is manifested by a shift in the intertropical convergence or meteorological equator (see p. 71), whose position on land separates the dry regions where the trade winds prevail from the humid regions where the monsoons blow (Petit-Maire, Sanlaville, and Yan 1995; Leroux 1996).

Until the 1970s, many researchers argued against the idea of global warming. However, numerical data collected by satellites and treated by increasingly sophisticated mathematical models has shown that globally the climate is growing warmer, while also showing greater contrasts. At any rate, despite a few persistent skeptics, the GIEC/IPCC consider global warming to be a real phenomenon. Over the past 150 years, temperatures have risen by 0.3°C–0.6°C and could increase by 2°C–5°C by the end of the century. The causes for this phenomenon as well its consequences are still bitterly debated. No natural cause, by itself, can explain this sudden acceleration in warming, not even the sporadic diffusion of volcanic aerosols or desert dust, which would play a role in reflecting back terrestrial radiation. We are truly in the midst of a climate disruption of a new kind, as human action is undoubtedly responsible for it, particularly through the proliferation of chemical pollution in the form of "greenhouse gases" (CO_2, N_2O, CH_4) and gases that destroy the ozone layer (CFCs[5]). While absent or negligible in earlier times, human contribution to climate change is a formidable factor today (although we must also recognize its limits) (Leroux 1996; Godard 2001).

Scarcely populated and barely industrialized, deserts cannot be considered a cause for global warming. In fact, they risk suffering the consequences of global warming far more severely than temperate regions. The possible threats they could face explain why researchers have developed scenarios around future climatic changes. The future

5 CFCs: chloro-fluoro-carbons.

is imagined from the assumed or probable condition of present-day variables, "tuning" mathematical models on a global scale. This does pose problems, and sometimes yields contradictory results which are dubious, to say the least. However good the quality of the models, how can we be certain of what the population will be like in 50 or 100 years? What activities people will perform and the pollutants this will produce? When considering global variations, how do we take into account the primordial role of local and seasonal variations, such as the start date or the duration of the rainy season? How can we gauge how effective oceans and forests will be as "carbon sinks" in removing gaseous pollutants? Or, again, what can be done about the "hole" in the protective ozone layer?

The scenario discussed most often is that of an interglacial situation evolving, over thousands of years, into a glacial future, but severely disrupted by anthropic-driven warming, accelerated over the last decades (Labeyrie 1993; Joussaume 1999). This model must, of course, include all possible nuances related to local geographic conditions. Global warming is likely to lead to a pluvial climate (see section 3.5.2.2, p. 129), but this does not exclude the occasional occurrence of severe droughts. However, cloud cover of increasing thickness and a lack of sunlight in the higher latitudes will lead to temperatures dropping and an interpluvial-type climate being established. After several oscillations between these two extremes, marked by the simultaneous displacement of the intertropical convergence of the trade winds and monsoons (Petit-Maire, Sanlaville, and Yan 1994; Petit-Maire, Sanlaville, and Yan 1995), an appropriate orbital forcing will re-establish glacial conditions, with the maximum extension of deserts across the Earth's surface.

3.6. Unusual phenomena caused by or promoted by aridity

Almost everyone who has traveled in the deserts has, at some point or the other, encountered strange, fleeting phenomena that are often spectacular, strange or troubling. The aridity leads to physical conditions that are conducive to the development of meteors in the atmosphere that are unknown or rare in other climatic regions. Some of the most remarkable of these phenomena are discussed in this section.

When you have toiled an entire day on foot, camel or a bumpy car seat, you come to be deeply appreciative of being able to relax in a camp in the evenings. And as you fall asleep, either stretched out on a camp bed or, more often, in the field, under a starry sky, you sink into the solitude and the silence, the silence of these regions without water and without birds, the "eternal silence of the infinite spaces", that frightened Pascal. In the day there are always the voices of your companions, the cries of animals or the sound of engines and the disturbances in the environment. But in the cold and dense night air, all the space around you is weighed down by silence. Sounds are rare and unusual:

the wind rustling through the bushes of broom or saxaul, or straying through a valley or the dunes, the muffled sound of a caravan marching through the night, the wild cry of a rutting camel, the distant call of a predator... sounds carry with astonishing clarity and over considerable distance in this bare and treeless land. And when the isolated travelers finds themselves alone with their thoughts, their peace or their own anguish, it is a moment for them to "plunge into this silence as they journey through the desert", being not afraid, but humble (T. Monod).

Another charm that desert nights offer is the evening cool and the splendor of the star-studded sky. There is always a perceptible drop in temperature as soon as the Sun goes down and after the 35 or 40°C of the day, the night-time temperatures of 18 or 20°C are quite cold. But the relaxed evenings are always soothing. Moonlit nights are so bright that it is possible to read without a lamp, a blinding light blocks out the stars and gives the naked rocks a ghostly appearance, casting deep shadows. But on dark nights, through the dry and clear skies, the stars appear infinitely more numerous, larger and brighter than they do anywhere else. You cannot tire of looking at the stars travel across the skies, watching entire constellations rise and set. In the early hours of the morning, just before daybreak, the temperature, which has reduced following the nocturnal radiation, is at its lowest. Even in tropical deserts, in winter, it can be freezing cold and frost lies on all objects and on the soil. And then there comes the luminous burst of light that signals the day, sudden and rapidly growing brighter. The heat rises, growing every minute, and at the same time, all colors fade into a uniform and omnipresent monochrome yellow. The heat glues the walkers to the ground and frequent breaks are necessary. There are no more cool shadows; even in the shade of rocks one is subject to the heat radiated by the rocks; the shadow cast by the rare tree shrinks rapidly as a camel snacks on the leaves. You set off again in the afternoon, when the heat has reduced a little. Twilight is as rapid as daybreak, but even more splendid. The suspended dusts, shot through by the rays of the setting Sun, spread reds and flamboyant violets over the ochres of the relief, which are then sucked away by the shadows of the night.

The presence of fine, suspended solid particles in the air (aerosols) is a near-permanent feature of the desert atmosphere. This terrigenous pollution results from the wind winnowing the surface of the soil that is bare or that has been degraded by human activities. Made up of fine grains (clay, silt or sand), visible and highly mobile, this pollution makes up lithometeors, which are characteristic of arid and semi-arid regions (Nouaceur 2002). They have become even more frequent over the past few decades, in parallel with the increased frequency of droughts and growing desertification. These lithometeors take on several shapes, increasing or decreasing in intensity depending on the temperature and humidity in the air, the force of the wind, the granulometry of the suspended particles and the configuration of the relief over which they blow. Among the most frequent and widespread forms are dry mists, whirlwinds, dust-storms and sand-storms.

Dry mists are aerosols formed by microparticles carried by rising heat and scattered through air that has very little water vapor. They appear like a veil that is more or less dense and sifts and diffuses sunlight. This often gives the skies over hot deserts the yellowish tint that is so different from the transparent luminosity of polar skies in good weather.

Sometimes, during the day, *whirlwinds* of dust can form over plain and denuded surfaces that have been unequally warmed by the Sun (due to differences in light received, in colors and in albedo). These miniature tornadoes (see section 3.2.1.5, p. 71 *sqq.*) are extremely mobile and called *dust devils*, *dust vortexes* or *djnoon* (the Arabic plural for djinn). The thermal instability in the atmospheric layer closest to the ground causes overheated air to rise (convection) and form a sort of chimney, and the dust that was raised starts spinning around this. Whirlwinds have been recorded spinning at speeds of 20–40 m/s, which is high enough to carry even sandy particles 50–200 µm in size. The whirlwind travels along a trajectory described by the heated points in the field, conserving angular velocity and its direction of rotation. A whirlwind can last for a few minutes to even half an hour, depending on its diameter and the height to which it grows (they can reach heights of up to 300 m, sometimes even more). Whirlwinds are frequent in summer, in dry seasons and in the early afternoon, when convection is most intense. They fall under local winds when air movement is more active and the temperature of the air falls as a result. Whirlwinds are chiefly composed of fine sediments, silt, loess and dried clay, in lake-beds or sebkhas, and in the bed of dry wadis that orient their movement.

Dust storms and *sand storms* are phenomena in which a mass of loose materials, usually lutites (< 50 µm) or arenites (from 50 µm to 2 mm), is transported in a turbulent air current. Like whirlwinds, both these phenomena are promoted by the denudation of arid soil, which is almost bare of vegetation. However, unlike whirlwinds, their movement is guided by local or regional meteorological pressure fields. Most of them occur between the end of winter and the beginning of summer, when general atmospheric circulation is most intense.

In the case of *dust storms*, the vertical movement of air and dust can reach as high as 3,000 m–4,000 m above the ground, blocking out the sky, and the displacement (which can be tracked by satellites) can extend over several thousand kilometers. Certain Saharan dust storms have traveled the Mediterranean (due to the sirocco), and others have crossed the Atlantic as far as the Caribbean (Coudé-Gaussen 1991). When they arrive in a more humid atmosphere, the sand grains adsorb water molecules, become heavier and fall to the ground forming "blood rain" or "red rain", which has been described from Egypt to Rome since antiquity, and rain dust, which can sometimes even reach temperate environments.

Sand storms are more turbulent and more unpleasant for the traveler. Made up of very heavy grains, often quartz, they lash at faces, blow into bags and destroy food. Sometimes they are dense enough to obscure the sky, but they move close to the soil and are usually not very dense, only a few meters thick.

Dust storms and sand storms are very efficient agents in transporting loose material and, consequently, they play a prominent geodynamic and geomorphological role in the interior of deserts and along their periphery. During the glacial periods, dust storms, in particular, contributed to the thick, wind-borne loess sedimentation in Asia and Europe. However, they do not create any unusual shapes. Sand storms, on the other hand, as they move very close to the ground and are more affected by topography, have more varied effects, which are also more localized. They are capable of creating and sculpting original shapes – the classic example of this are the dunes. Their morphological role is examined in greater detail in the chapter on the wind (Volume 3, Chapter 1).

Mirages are another example of an unusual phenomenon in the desert. Frequently observed during hot days in the desert, they are optical illusions created by the refraction of light rays traveling through overheated layer of air (heated through contact with the ground) into the colder layers above them. An observer perceives a virtual, blurred, out-of-proportion image of an object that may seem to float above the soil or even be inverted. There are mirages of all sizes and all kinds of illusions that one can observe. The simplest of these are the appearance of a pool of water glinting in the sunlight, which is, however, simply the reflection of the sky in the distance. Other mirages often seem to be offering visions of vegetation or large buildings, deformed, as if waving in the wind, unnaturally raised or floating on an imaginary lake, all of which moves further away as you approach it. Some authors have tried to use mirages to explain famous mysteries such as the Red Sea opening up before the Hebrews, or Jesus walking on water. A mirage has even been used as an aesthetic object, as in the famous scene in David Lean's Lawrence of Arabia (1962), where one sees a camel driver come up over the shimmering horizon, trembling and wavering on the smooth surface of a sebkha. Despite its stunning appearance, it remains a banal physical phenomenon in overheated arid regions.

Storms are not specific to a desert, but a desert offers certain conditions that are highly conducive to their development. First of all, there is a large reserve of static electricity, chiefly due to the abundant superficial formations made up of isolated grains (sand) and, secondary to this, the presence of rocks rich or enriched with iron. This latent static electricity sometimes manifests itself in a tangible manner, e.g. irritable or aggressive behavior in individuals or animals before an approaching storm or sandstorm. It also manifests as a large vertical instability in the lower atmosphere. As air gets heated by contact with the denuded soil, the temperature differences between the warm air and the cooler upper layers get more exaggerated. At the same time, evaporation and heat carry the maximum absolute humidity (see p. 68). Convection movements that follow move vertically from layer to layer through a rapid chain of warming–rising–relaxing–condensing (see p. 70), over a section that is 2–4 km thick. Each time the air reaches saturation point, clouds form with a vertical development that release polarized water droplets and create – between the drops and the earth–the potential differences required to generate lightning. In the deserts, given the absence of water, storms are often dry storms, violent but without precipitation. They usually occur in hot weather, at the end of the day, in low winds and a stable pressure field. They can impact a large area. Sometimes, external circumstances can aggravate storms by worsening instability or increasing the humidity, for example, the cold fronts that accompany the migration of MPAs, or the orographic rising of warm air in the mountains or, again, the high humidity carried by the tropical monsoons. Storms release a large amount of energy, which then acts upon the geodynamic evolution of the Earth's surface. It is said that lightning is capable of destroying composite rocks like ferricretes or even granular rocks like granite or sandstone, or that it can fissure and melt exposed or isolated rocks. At any rate, it is responsible for the creation of *fulgurites*, which are widely scattered through dunes in a desert, and which are formed by the melting and then solidification of silica in sand to form tubular or branched glass. Some storms may also be accompanied by heavy rain, which streams away across the soil and causes large volumes of water to flow through talwegs. These flows are sudden and ephemeral.

It may seem paradoxical to talk about *fire* ravaging a land that is nearly empty of all habitation and bare of vegetation. However, the arid and sub-arid borders of a desert are not free of this danger, given the dry air and desiccated vegetation. Above all, the steppes and grassy savannas that border temperate and tropical deserts are most at risk, whether from accidental fires from lightning or deliberate fires started to stimulate fresh growth or to prepare new farmland.

3.7. Aridity and drought

Aridity and *drought* are indeed the fundamental characteristics of a desert. They arise from a combination of multiple causes that involve the atmosphere, the earth and even extra-terrestrial causes, on the one hand, while also involving the human condition, on the other. Variable across space, aridity offers geographic nuances that can be measured to varying degrees and that highlight the large diversity in deserts that people are often unaware of. A phenomenon that has existed from the earliest geological ages, aridity has often left sedimentological and geomorphological evidence which can often still be seen in certain landscapes. From the end of the Cenozoic period and during the Quaternary period, arid zones were the cradle of humanity and the environment in which humans first developed. By definition, arid zones lack water, although they cannot simply be reduced to a region with deficit rainfall. There are other remarkable properties that characterize them, such as the wind and the specific plant and animal life that make up arid landscapes and ecosystems, adding to the personality of a desert and accentuating the variety and complexity of these spaces.

3.8. References

Allègre, C. (1983). *L'écume de la Terre*. Paris: Fayard. 366 pp.

Alsharhan, A., Glennie, W., Whittle, G. and Kendall C. (1998). *Quaternary deserts and climatic change. Proceedings of the international conference on quaternary deserts and climatic change*. Rotterdam: Balkema. XIV+622.

Arnaud-Fassetta, G. (2008). "Les terrasses fluviales." In *LES FORMATIONS SUPERFICIELLES, Genèse — Typologie — Classification — Paysages et environnement — Ressources et risques*. Ed. by Y. Dewolf and G. Bourrié. Paris: Ellipses, pp. 236–243.

Aubouin, J. (1975). "Stratigraphie." In J. Aubouin, R. Brousse, and J.-P. Lehman. *Précis de géologie*. Vol. 2. Paris: Dunod-Bordas, pp. 297–666.

Aubreville, A. (1949). *Climats, forêts et désertification de l'Afrique tropicale*. Paris: Société d'éditions géographiques, maritimes et coloniales. 351 pp.

Balout, L. (1952). "Pluviaux interglaciaires et préhistoire saharienne." In *Travaux de l'Institut de Recherches sahariennes*, 8, pp. 9–21.

Bardinet, C. (1994). "Télédétection multisatellite de l'aridité — Le cas des limites du Sahara et du Sahel africains." In *Milieux arides et semi-arides. Colloque en l'honneur du Prof. Coque* (ENS Saint-Cloud). Paris, pp. 85–109.

Brahic, A., Hoffert, M., Schaaf, A. and Tardy M. (1998). *Sciences de la Terre et de l'Univers*. Paris: Vuibert. 634 pp.

Budyko, M. (1958). *The heat balance of the Earth's surface*.Report. Washington, D.C.: U.S. Department of Commerce. 253 pp.

Capot-Rey, R. (1951). "Une carte de l'indice d'aridité au Sahara français." In *Bulletin de l'Association des Géographes français*, pp. 73–76.

— (1952). "Les limites du Sahara français." In *Travaux de l'Institut de Recherches sahariennes*, 8, pp. 23–48.

— (1953). *Le Sahara français*. Paris: Presses Universitaires de France. 564 pp.

Cosandey, C. and Robinson M. (2000). *Hydrologie continentale*. Paris: Armand Colin. 360 pp.

Coudé-Gaussen, G. (1991). *Les poussières sahariennes — Cycle sédimentaire et place dans les environnements et paléoenvironnements désertiques*. Paris: J. Libbey, Eurotext. 498 pp.

Demangeot, J. and Bernus, E. (2001). *Les milieux désertiques*. Paris: Armand Colin.

Dresch, J. (1982). *Géographie des régions arides*. Collection Le Géographe. Paris: Presses Universitaires de France. 277 pp.

Dubief, J. (1950). "Évaporation et coefficients climatiques au Sahara." In *Travaux de l'Institut de Recherches sahariennes,* 7, pp. 13–44.

— (1951). "Alizé, harmattan et vents étésiens." In *Travaux de l'Institut de Recherches sahariennes,* 7, pp. 187–189.

— (1959–1963). *Le climat du Sahara.* 2 vols. Mémoire hors série. Alger: Institut de recherches sahariennes. 312 & 275.

Emberger, L. (1930). "Sur une formule climatique applicable en géographie botanique." In *Comptes Rendus de l'Académie des Sciences,* 191, pp. 389–391.

— (1938). "La définition phytogéographique du climat désertique." In *Mémoires de la Société de Biogéographie* VI, pp. 9–14.

Flint, R. (1971). *Glacial and Quaternary Geology.* New York: John Wiley & Sons Ltd. 292 pp.

Frakes, L. (1979). *Climates throughout geological time.* Amsterdam: Elsevier. 310 pp.

Gardner, R. (1994). "Loess, climate and orogenesis, implications of South Asian loesses." In *Zeitschrift für Geomorphologie,* 38, pp. 169–184.

Gaussen, H. (1954). *Géographie des plantes.* Paris: Armand Colin. 128 pp.

— (1957). "Les climats biologiques et leur classification." In *Annales de Géographie,* pp. 193–220.

Gaussen, H. and Bagnouls, F. (1952). "L'indice xérothermique." In *Bulletin de l'Association des Géographes français,* pp. 10–16.

Gladfelter, B. (2000). "The geomorphic context of the Upper Paleolithic in Wadi Gayifa, northeastern Sinaï." In *Zeitschrift für Geomorphologie,* pp. 1–31.

Glantz, M. (1987). "La sécheresse en Afrique." In *Pour la Science* (August 1987), pp. 18–25.

Godard, A. (2001). "Changement climatique et effet de serre additionnel d'origine anthropique : un débat parfois obscur." In *Annales de Géographie,* 617, pp. 79–89.

Hobbs, J., Lindesay, J., and Bridgman H. (1998). *Climates of the South Continents — Present, past and future.* New York: John Wiley and Sons. 297 pp.

Houghton, J. et al. (1990). *Climatic Change, the IPCC scientific assessment.* Cambridge, University Press.

Joly, F. (1953). "Essai d'une définition du Présahara marocain." In *Bulletin de la Société Royale de Géographie d'Égypte* 25, pp. 57–70.

— (1994). "Le Présahara, marge écologique et pôle de développement." In *Milieux arides et semi-arides. Colloque en l'honneur du Prof. Coque.* Paris, pp. 405–415.

Joussaume, S. (1999). *Climat d'hier à demain.* Paris: CNRS. 143 pp.

Köppen, W. (1931). *Grundrisse der Klimakunde.* Berlin: Walter de Gruyter.

Köppen, W. and Geiger, R. eds. (1930–1939). *Handbuch der Klimatologie.* 5 vols. Berlin: Verlag Gebrüder Borntraeger.

Labeyrie, J. (1993). *L'homme et le climat.* Collection Points Sciences. Paris: Denoël. 349 pp.

Lamarre, D. and Pagney P. (1999). *Climats et sociétés.* Paris: Armand Colin. 272 pp.

Lamb, H. (1977). *Climate, present, past and future.* London: Methuen & Co. 835 pp.

Le Floch, E., Grouzis, M.,Cornet, A.,and Bille, J.eds. (1992). *L'aridité, une contrainte au développement*. I.D.T. Paris: ORSTOM. 597 pp.
Le Roy Ladurie, E. (1967). *Histoire du climat depuis l'an mil*. Paris: Flammarion. 319 pp.
Le Treut, H. and Jancovici J. (2001). *L'effet de serre*. Dominos. Paris: Flammarion. 128 pp.
Leroux, M. (1996). *La dynamique du temps et du climat*. Paris: Masson. 310 pp.
— (2002). "Global Warming, mythe ou réalité ? L'évolution réelle de la dynamique du temps." In *Annales de Géographie*, 624, pp. 115–137.
Lhote, H. (1968). *Art préhistorique*. In *Encyclopaedia Universalis*, 13. Paris, pp. 500–507.
Liu, T. ed. (1991). *Loess environment and Global Change*. Beijing: Science Press. 287 pp.
Lombard, A. (1972). *Séries sédimentaires — Genèse, évolution*. Paris: Masson. 425 pp.
Mainguet, M. (1995). *L'homme et la sécheresse*. Paris: Masson. 335 pp.
de Martonne, E. (1926)."Aréisme et indice d'aridité ". In *Comptes Rendus de l'Académie des Sciences*, 182, pp. 1395–1398.
— (1942). "Nouvelle carte mondiale de l'indice d'aridité." In *Annales de Géographie*, pp. 241–250.
Meigs, P., ed. (1977–1979). *Carte de la répartition mondiale des régions arides, Carte à 1/25 000 000, 1 feuille en couleurs, Notice explicative*. Notes techniques Man and Biosphere 7. Paris: UNESCO. 55 pp.
Miskovski, J.-C. (1988). *Les sédiments, témoins du passé*. Monaco: Éditions du Rocher. 122 pp.
Nesme-Ribes, E. and Thuillier G. (2000). *Histoire solaire et climatique*. Paris: Belin. 240 pp.
Nicholson, S. (1978). "Climatic variations in the Sahel and other African regions during the past five centuries." In *Journal of arid environments*, pp. 3–24.
Nouaceur, Z. (2002). "Les lithométéores : des types de temps caractéristiques des régions arides et semi-arides." In *Sécheresse*, 13 (2), pp. 87–93.
Pedelaborde, P. (1970). *Les moussons*. Paris: Armand Colin. 221 pp.
Penman, H. (1956). "Evaporation : an introductory survey." In *Netherland Journal of Agricultural Sciences*, 42, pp. 9–29.
Petit, J.R. et al. (1999). "Climate and atmospheric history of the past 420 000 dates from the Vostok ice core, Antarctica." In *Nature*, 399, pp. 428–436.
Petit-Maire, N. (1984)."Le Sahara, de la steppe au désert." In *La Recherche*, 160, pp. 1372–1382.
— (1999). "Carte des environnements du monde pendant les deux derniers extrêmes climatiques, 2 feuilles à l'échelle de 1/25 000 000." In *Comptes Rendus de l'Académie des Sciences, Paris, Sciences de la Terre et des Planètes*. II 328 (4): *L'effet de serre, certitudes et incertitudes*. Ed. by M. Petit, pp. 221–293.

Petit-Maire, N., Sanlaville, P., and Yan, Z. (1994). "Changements globaux et paléomoussons : la zone de transition afro-asiatique au cours des derniers 140 000 ans." In *Géochronique,* 50, pp. 16–19.

— (1995). "Oscillations de la limite nord du domaine des moussons africaine, indienne et asiatique au cours du dernier cycle climatique." In *BSGF,* 166 (2), pp. 213–220.

Raynal, R. (1972a). "Analyse des caractères originaux des divers secteurs morphoclimatiques sahariens." In *Colloque sur les déserts*. Paris: Bulletin de l'Association des Géographes français, pp. 196–201.

— (1972b). "Essai de définition de trois degrés dans l'aridité désertique." In *La pensée géographique contemporaine. Mélanges A. Meynier*. Rennes: Presses Universitaires de Bretagne, pp. 291–297.

Rognon, P. (1989). *Biographie d'un désert*. Paris: Plon. 347 pp.

— (1996). "Climatic change in the African deserts between 130 000 and 10 000 yrs BP." In *CRAS*. IIa 323, pp. 549–561.

Sanlaville, P. (1995). "Marge septentrionale du désert arabique la zone sud-levantine au Pléistocène supérieur et à l'Holocène." In *Bulletin de la Société Géologique de France,* 167, pp. 57–66.

— (2000). *Le Moyen-Orient arabe. Le milieu et l'Homme*. Collection U. Paris: Armand Colin. 264 pp.

Sécheresse, ed. (1995). Vol. 6. Agence Universitaire de la Francophonie, John Libbey Eurotext, pp. 3–158.

Siegert, M., Ruecker, G., Hinrichs, A., and Hoffmann A. (2001). "Increased damages from fires in logged forests during droughts caused by El Ni no." In *Nature,* 414 (Nov.), pp. 47–440.

Smolka, P. and Volkheimer, W. eds. (2000). *Southern Hemisphere. Paleo and Neoclimates*. Berlin: Springer Verlag. 381 pp.

Tardy, Y. (1986). *Le cycle de l'eau — Climats, paléoclimats et géochimie globale*. Paris: Masson. 338 pp.

Tardy, Y. and Roquin, C. (1998). *Dérive des continents — Paléoclimats et altérations tropicales*. Orléans: BRGM. 473 pp.

Termier, H. and Termier, G. (1979). *Histoire de la Terre*. Paris: Presses Universitaires de France. 430 pp.

Thornthwaite, C. (1948). "An approach toward a rational classification of climate." In *Geographical Review*, pp. 55–94.

Turc, L. (1955). "Le bilan de l'eau des sols, relation entre les précipitations, l'évaporation et l'écoulement." In *Annales agronomiques,* 6 (1), pp. 5–152.

— (1961). "Évaluation des besoins en eau d'irrigation — Formule climatique simplifiée et mise à jour." In *Annales agronomiques,* 12 (1), pp. 13–49.

Vigneau, J. (2001). "La zone « supratropicale» dans le monde et en France — Réflexion sur la classification des climats." In *Annales de Géographie,* 619, pp. 227–242.

Voituriez, B. and Jacque, G. (1999). *El Niño, réalité et fiction*. Collection COI Forums Océans. Paris: UNESCO. 140 pp.

Wegener, A. (1915). *Die Entstehung der Kontinente und Ozeane.* La genèse des continents et des océans. Trans. by A. Lerner. Paris: Nizet et Bastard. 236 pp.

Wilson, R., Drury, S., and Chapman J. (2000). *The great Ice age : climate change and life.* London: The Open University. 257 pp.

WMO (1966). *Climatic Change.* Technical Note 79. Geneva: World Meteorological Organization.

Wünnemann, B., Pachur, H., and Zang, H. (1998). "Climatic and environmental changes in deserts of Inner Mongolia, China, since the late Pleistocene." In A. Alsharhan, W. Glennie, G. Whittle, and C. Kendall. *Quaternary deserts and climatic change. Proceedings of the international conference on quaternary deserts and climatic change.* Rotterdam: Balkema, pp. 381–394.

List of Authors

Yvette DEWOLF
Paris Diderot University – Paris 7
France

Fernand JOLY (†)
Paris Diderot University – Paris 7
France

Index

M, N

P

R, S

T, U

V, W, X

Other titles from

iSTE

in

Ecological Science

2020

BRUSLÉ Jacques, QUIGNARD Jean-Pierre
Fish Behavior 1: Eco-ethology
Fish Behavior 2: Ethophysiology

ROSSIGNOL Jean-Yves
Climatic Impact of Activities: Methodological Guide for Analysis and Action

2019

AMIARD Jean-Claude
Industrial and Medical Nuclear Accidents: Environmental, Ecological, Health and Socio-economic Consequences
(Radioactive Risk SET – Volume 2)
Nuclear Accidents: Prevention and Management of an Accidental Crisis
(Radioactive Risk SET – Volume 3)

BOULEAU Gabrielle
Politicization of Ecological Issues: From Environmental Forms to Environmental Motives

GIRAULT Yves
UNESCO Global Geoparks: Tension Between Territorial Development and Heritage Enhancement

KARA Mohamed Hichem, QUIGNARD Jean-Pierre
Fishes in Lagoons and Estuaries in the Mediterranean 2: Sedentary Fish
Fishes in Lagoons and Estuaries in the Mediterranean 3A: Migratory Fish
Fishes in Lagoons and Estuaries in the Mediterranean 3B: Migratory Fish

OUVRARD Benjamin, STENGER Anne
Incentives and Environmental Policies: From Theory to Empirical Novelties

2018

AMIARD Jean-Claude
Military Nuclear Accidents: Environmental, Ecological, Health and Socio-economic Consequences
(Radioactive Risk SET – Volume 1)

FLIPO Fabrice
The Coming Authoritarian Ecology

GUILLOUX Bleuenn
Marine Genetic Resources, R&D and the Law 1: Complex Objects of Use

KARA Mohamed Hichem, QUIGNARD Jean-Pierre
Fishes in Lagoons and Estuaries in the Mediterranean 1: Diversity, Bioecology and Exploitation

2016

BAGNÈRES Anne-Geneviève, HOSSAERT-MCKEY Martine
Chemical Ecology

2014

DE LARMINAT Philippe
Climate Change: Identification and Projections

Printed in the USA/Agawam, MA
September 16, 2026

910661.018